Tuning for Economy and Performance

By David Rowlands
and the Autobooks team of Technical Writers

Autobooks

Autobooks Ltd. Golden Lane Brighton BN1 2QJ England

Acknowledgements

The Publishers wish to thank the following firms for their assistance in the preparation of this book:

Aeon Products (London) Limited

Armstrong Patents Company Limited

Associated Tyre Specialists Limited

Automotive Products Group

J. W. E. Banks & Son Limited

Brown & Geeson Limited

Leslie Hartridge Limited

Leonard Reece & Company Limited

Joseph Lucas Limited

Lumenition Limited

S.P.Q.R. Engineering Limited

Tapley Meters Limited

Wolfrace Wheels

Cover Photography: Walter Gardiner

First Published 1975

© Autobooks Ltd 1975

ISBN 0 85147 577 9

650

Printed and bound in Brighton England for Autobooks Ltd by G. Beard & Son Ltd

Contents

Foreword   8

Chapter 1   What tuning means   9

Chapter 2   Work on the engine   21

Chapter 3   Tuning the ignition system   49

Chapter 4   Transmission   67

Chapter 5   Tuning the suspension   77

Chapter 6   Wheels, tyres and brakes   89

Chapter 7   Bodywork, accessories and instruments   99

Chapter 8   Legal requirements and insurance   103

Chapter 9   Tuning and new car warranties   107

Chapter 10   Testing a tuned car   109

Foreword

A great deal is heard in motoring circles and written in the motoring press about tuning, and many motorists, already perhaps carrying out their own car maintenance and servicing at home, will have wondered precisely what is meant by tuning and what it can offer them. Clearly it is no longer the exclusive province of the professional specialist – the wealth of tuning aids on sale in garages, accessory stores and speed shops throughout the country is evidence of that. Equally it is no longer solely the concern of motor sport enthusiasts and those seeking high performance at all costs, as more and more people are realising that by enhancing the car's efficiency tuning can provide more economical motoring. This book introduces the reader to the subject, offering a definition of tuning, a survey of the methods and operations involved and an assessment of the benefits they can bring. In this way it aims to help the would-be tuner to define his or her objectives and then to provide practical advice on how those objectives can be achieved. In addition to the measurable gains of money saved or performance improved there is the satisfaction of developing one's skill and applying it successfully. For whatever the lack of specialised equipment and advanced technical knowledge, the amateur has one great advantage – that is time, time which no mass-production manufacturer can afford to expend on any individual vehicle. The pages that follow explain some of the best ways to use that advantage, as well as some of the pitfalls to be avoided.

David Rowlands

CHAPTER 1

What tuning means

1:1 The scope for tuning
1:2 What kind of operations are involved in tuning?

1:3 A balance sheet for economy tuners
1:4 Good workshop practice
1:5 Tools, instruments and their care

1:1 The scope for tuning

Mass production cars are relatively imperfect machines. They are made at high speed by men with no time to take an interest in the vehicle or its future owner. The men use parts poured out in millions from smaller engineering works that are themselves only responsible for a minute part of the total vehicle. The parts from one small factory are expected to match those from another. They don't, of course; but they fit well enough for there to be little apparent fault in the car or its performance.

The car is not only imperfect – its design is relatively crude. So crude in fact that it can accept these imperfections in the fit of its parts with indifference. Simplicity or crudity are deliberate ploys by the manufacturer to limit the car's performance and thus enable the car's engine and other stressed components to perform with some semblance of reliability, mile after mile seemingly developing the same power output, handling and behaving the same. Modern cars will put up with relatively high engine speeds and poor treatment from the driver for very long distances. Engines, for instance, typically run for 50,000-120,000 trouble-free miles with nothing more than regular maintenance.

All these factors are behind the design of a mass production car. As the ultimate competitive industrial consumer product the car has to be manufactured as cheaply as possible and it has to work reliably for a well-defined time scale. Parts are cheapest in large quantities – the less restrictions on their dimensions and quality of finish the car manufacturer imposes on his suppliers the cheaper the parts become. By deliberately placing restrictions on the car's performance the manufacturer ensures his product is tolerant enough to accept mismatched parts and that it will accept the abuse of a car's normal life with some reserve of strength to cope with the most excessive demands of the owner. VW Beetle engines are renowned for lasting up to and over 200,000 miles – but then they are also known as perhaps the most restricted engines in modern production cars.

During the service life of the car parts wear, corrode, fatigue and break. They are replaced or adjusted at intervals to maintain the car's condition and its original performance.

Unfortunately most drivers are poor judges of a car's gradually changing performance characteristics. Day by day the deterioration is slight, so it escapes the driver's notice. The driver is a great compensator for the car's inadequacies. As the engine loses power he presses harder on the accelerator to regain performance at the expense of fuel. Cars can be using up to 15 per cent more fuel than should be required to produce a given power output without showing any signs of rough running or other symptoms that would nudge the driver into going for a service.

Herein lies one aspect of tuning – keeping the parts of a car up to the manufacturer's specifications. Pure tuning consists of the regular maintenance of items like the carburetter, ignition system, brakes and tyres. But, the design of cars also allows considerable leeway for the tuner to improve on the manufacturer's specifications and this is the second aspect of tuning. The tuner can do all those jobs that the car's manufacturer could not possibly afford to do, like matching parts to very close tolerances and improving the finishing of engine components.

There is a third aspect to tuning which has become a prominent industry in the last twenty years or so. It is the replacement of the cheap mass-produced components of a car with modified or specially manufactured parts having greatly improved characteristics.

A useful definition of tuning, covering all three of these processes, is that it is the reduction or removal of the constraints on the process of energy transformation that takes place in the combustion chamber and the eradication of factors which diminish the resulting power output between the engine and the point where the road wheels meet the road.

It follows from this definition that tuning holds out two potential benefits to the motorist. Removing constraints on the combustion process and so increasing its efficiency, makes better utilisation of the energy stored in the vehicle's fuel. Equally, reducing those factors which restrict the amount of power available at the road wheels contributes to efficiency in the process of turning petrol into motive force. So the result of tuning for the motorist can be more miles per gallon.

But just as the extra power available from a given volume of fuel can be used sparingly to travel further it can also be used to accelerate faster. In other words the twin benefits of tuning are economy and performance.

These benefits are, however, to a large extent mutually exclusive. You can have one or the other, rarely both. A tuned car specifically designed for economic fuel consumption will usually have better performance characteristics than a similar car in an untuned state. If this extra margin of performance is used to any extent the economic benefits accruing from the design will vanish almost completely. In fact if the two cars were driven in such a way that there was no fuel saving (the overall fuel consumption of each car was the same) the only remaining benefit would be that the tuned car arrived a few seconds before the untuned car, everywhere they went.

Since a few seconds advantage over an untuned car is not really worth having and speed limits have reduced the possibility of travelling at speed for long distances, tuning for performance is often linked to some branch of motor sport such as rallying, autocross or racing.

Far more people are becoming interested in tuning as a means of reducing fuel bills as petrol prices soar and other costs follow similar inflationary trends.

Whereas the regular tuning tasks to compensate for wear and so on, like a great deal of the removal of the engine's imperfections, can be carried out at little expense to the owner, bolting on race and rally equipment (and particularly so-called economy devices) can be a very pricey occupation. Clearly if the car owner is expecting to obtain economic running from his car, any outlay on parts or specialised operations must be more than compensated for by the returns from petrol savings within the useful lifetime of the car. If performance alone is the aim the spending can go on for as long as the pocket will allow in the quest for perfection.

In the following sections some guide to the costs of tuning processes, parts and tools are given. Using the methods discussed below it is possible to work out a tuning balance sheet which will serve as a way of keeping in pocket for economy tuners and inform performance tuners of the escalating cost possibilities.

1 : 2 What kind of operations are involved in tuning?

As explained above there are three distinct types of tuning process. In its purest form, tuning is the adjustment of carburetter, ignition, brakes, suspension, steering, and so on to achieve the best possible performance, either for speed or economy. Many of these operations will form a part of normal garage or home servicing although there is a great deal of evidence to show that the tasks are all too frequently not carried out often enough or well enough to make a great deal of difference to the car's performance.

Certainly there are temptations to make components such as plugs and contact breakers soldier on long past the time when they can operate most efficiently. To do so is a false economy. Few motorists check tyre pressures as often as they should; under-inflation increases rolling resistance and raises fuel consumption and over-inflated tyres wear faster. Binding brakes will also take their toll of the fuel consumption.

Few routine tuning operations for the majority of cars are beyond the scope of the amateur with little automobile experience. Taking the time and trouble to study the operations and carrying them out with care not only saves the cost of garage servicing (perhaps as much as £15 three times a year) but also can make tremendous fuel savings and reduce the number of spare parts the engine requires throughout its useful life.

Removing the imperfections inherent in the design of a car can be a more complex task but once the various operations involved have been carried out, a once and for all gain will be made in improved performance and lowered petrol consumption.

Poor finishing and ill-fitting components in the engine are major causes of restricted power. Surface imperfections in the ports, manifolds and combustion chambers introduce irregularities in the flow of the petrol air-mixture resulting in imbalance in the amounts of combustion mixture in each cylinder and can cause irregular combustion. The result is inefficiency in the combustion process.

Where cylinder head ports and manifolds are joined the wide tolerances which the car manufacturers allow to their components suppliers cause considerable overlapping. This gives rise to further restrictions or irregularities in the gas flow. Even the cylinder block and the combustion chambers in the cylinder head may be poorly matched in relation to one another. Ironing out these faults on the production line would increase the cost of the car by a very large margin. Improving gas flow and matching the car's standard components costs the home tuner very little money; an inexpensive set of special tools and a lot of time are the primary essentials.

The ways in which the car's performance is deliberately restricted to ensure the reliability of crudely machined, balanced or cast components like the crankshaft and connecting rods should also be examined during the gas-flowing and matching process. Some of the restrictions to power output can be removed by the same techniques as are employed for matching components – typically the induction manifold and cylinder head ports can be opened out. However, cleaning up many of the restrictions can only be achieved by replacement of the car's standard parts.

For some cars it may be possible to find parts from a high performance car in the same model range at a scrapyard, through the Motor Vehicle Dismantlers Association or by advertisements in specialist publications. Parts like inlet and exhaust manifolds of a better, less constricting design can sometimes be obtained quite cheaply. But as cylinder heads and other basic performance parts are in considerable demand scrapyards may prefer to sell the whole engine.

Often the greatest restriction to power output on a standard saloon is the carburetter. A larger carburetter or a pair of carburetters might be obtained from a scrapyard but there are considerable pitfalls to successful fitting. Carburetters wear much more than people usually imagine and a cheap unit almost certainly means taking one with a considerable mileage behind it. Results may be very poor compared to those obtainable using a new carburetter of a larger or higher performance design.

This brings tuning into the very expensive area of bolt-on performance parts. Major manufacturers are aware of the demand that exists to make their standard models into rorty sports-car-eaters and they have not been slow to bring their own ranges of specially designed parts onto the market. Specialist tuning firms have similar ranges of bolt-on parts for production saloons.

The kind of parts available from these sources include ready-modified cylinder heads, performance camshafts, balanced and toughened crankshafts, high performance pistons, specially designed exhaust manifolds and oil coolers. The engine is not the only part of the car given serious treatment. Transmission, suspension and wheels have an equally long list of potential modifications.

Naturally a law of diminishing returns operates as tuning becomes more complex. The first 20 per cent increase in engine power is very easily obtained simply by rectifying the manufacturers production faults and ensuring that the car's systems are correctly adjusted. This is by no means an uncommon power increase for this state of tune – it may be accompanied by fuel consumption improvements in the order of 10 per cent (if the power is not utilised) and certainly the car will be more pleasant to drive. The engine will be more flexible, running evenly and quietly.

But successive tuning stages will cost more money and result in smaller gains in power and tiny improvements in fuel consumption. It is not an improbable task to double the power output from a standard engine. However, there would not be very much of the original engine left – major parts like the pistons, cylinders head and camshaft would all have to be replaced at great expense.

Between these two extremes there is a state of tune that will suit practically everybody's motoring needs. It is usually desirable from a cost point of view to define the state of tune, relating it as nearly as possible to the type of motoring for which the car is required, at the outset of a tuning programme. It is rarely convenient for the home tuner to dismantle an engine more than once or twice in the average life of the vehicle with any one owner. So it is a good plan to carry out any crankshaft and connecting rod balancing at the same time as head and manifold polishing.

What are the tuning possibilities to match various motoring requirements? The following are suggested tuning programmes to create cars of various temperaments:

1 For a standard car in economy tune:

Adjust carburetter to manufacturer's specification or use a Colortune instrument (see **Chapter 2**) to make fine adjustments so that the mixture is finely poised between rich and lean. Set contact breakers accurately and replace every 5000 to 6000 miles. Adjust timing stroboscopically to manufacturer's recommendation – slight retardation may sometimes allow use of one grade lower petrol (see later chapters for timing instructions and tests). Change plugs at least every 10,000 miles. In winter, fit a winter thermostat (high temperature) and/or blank off all or part of the front grille. Use radial tyres, accurately inflated to the maximum pressure recommended by the car manufacturers. Ensure brakes are not binding.

2 Mild performance increase—fun car or increased economy:

One size larger carburetter and manifold or a change to a carburetter with better characteristics plus all adjustments in **1**. To improve engine temperature and reduce noise and power loss consider fitting a feathering or electric fan replacement for standard fan.

Carry out head polishing and cleaning out of manifold to match head ports. Balance combustion chamber volumes and skim head to restore compression ratio. Polish valves and reduce valve seats.

3 Tractable road car:

Suitable for club rallying and offering improved economy if the power increase is not used. Carry out all operations as in **1** and **2**. In addition, fit flowed exhaust manifold. Fit a mild rally or sport camshaft, skim head to increase compression ratio. Fit high performance coil and performance distributor and/or contact breaker set (consider possibility of transistor assisted ignition). Consider necessity for heavy duty oil pump, and competition clutch.

At this state of tune the possibility is that the standard brakes will not cope with the extra performance. Examine the possibility of using different pad or shoe material or fitting disc brake conversion kit. The suspension will almost certainly not cope with full power – uprate shock absorbers, consider anti-roll bar or Panhard rod modifications to front and rear suspension. Lowering the suspension will be necessary at this stage. Aerodynamics are beyond the scope of most amateurs to design and perfect – take clues from the manufacturer's competition cars for spoiler and air dam body modification. Wider wheels will certainly be required to uprate road-holding and traction.

A car tuned to this specification will have lost few of its attractions as a road car – it will be extremely lively and of course a great deal of time and money will have been spent on it.

4 Race/rally car:

It is at this stage that any pretence of economy flies out of the window. From now on the tuning will become extremely expensive. The first pre-requisite of a full blown rally car that will not disgrace itself in track events will be a suitable cam with a high lift. This will certainly destroy the engine's tractability – combined with the necessary lightening and balancing of components like the flywheel, crankshaft and connection rod the effect is to make the engine lumpy at idling and slow speeds. The engine will be adjusted to bring maximum power in at 4000 to 6000 rev/min – it will barely function at slower speeds. Most other high stress components will have to be replaced with competition parts like rocker gear, duplex timing chains and sprockets, connecting rods, pistons, apart from the many adjustments in tuning state **3**. Every moving part will have to be minutely balanced and the car will have to be even more tightly strapped down onto its suspension. Transmission modifications may well become necessary.

Cars prepared for competition use have to conform to class and group regulations which lay down in great detail what must be done, what can be done and what cannot be done. In some cases the rules will virtually define the state of tune permissible. For some groups some of the measures suggested so far would be excluded, other groups are much freer.

5 Racing saloon:

Really full blown racing cars need a racing camshaft – one that is only suited to an engine revving smoothly within a very narrow range. Special heads with very large valves (sometimes an increase in the number of valves) will be used. Little of the production car's bodywork or engine will be left in the finished racer. Gaining precious fractions of a second in a race will rely on minute tweaks to the suspension, playing with the timing and carburation, adjusting the tyres used and a certain amount of luck to counteract the tendency of engines taken this far to blow up. This kind of preparation is out of the scope of this book – however, it is still possible for the racing enthusiast to pick up sufficient tips from following chapters to point the way towards the higher flown aspects of tuning.

1:3 A balance sheet for economy tuners

Ignoring the bottomless expenditure that can be indulged in to obtain high performance, it has been established that in serious economy tuning for everyday motoring the benefits from the tuning have to outweigh the expenditure by a large factor for the operation to be worthwhile.

In a balance sheet of the economy tuning process, what are the debits that must be taken into account? The list below is not exhaustive and in each case the costs are variable as different manufacturer's parts prices vary and differing quantities arise.

Direct costs:

Parts like contact breakers, plugs, gaskets.

Special tools that have no other function than in a single tuning operation – small grindstones, valve suction grinder, piston ring clamp.

Charges for engineering operations like cylinder head skimming, valve seat cutting and balancing.

Material costs – paraffin, antifreeze, gasket cement, Gunk, Swarfega.

Power for heating, lighting of workshop, electricity consumption of power tools.

Capital costs:

Tools cost amateur car mechanics a lot of money, often more than they realise. It would not be unreasonable to suppose that a decent set of tools for servicing and

tuning a car would cost £120 to £150. Assuming their use for major car servicing tuning and repair operations ten times a year for a period of, say, ten years, each operation incurs an element of tool costs in the region of £1.20 to £1.50.

Indirect costs:

There are always hidden costs involved in tuning a car. Examples are the extra costs of taking public transport when the car is stripped down and cleaning dirtied clothing.

The savings that can be expected:

It is never possible to place an exact figure on the sort of savings that can be expected in advance of an economy tuning operation – the gains have to be totted up afterwards by means of careful testing and calculations. The table below refers to cars travelling 10,000 miles a year – although the national average mileage travelled is slightly below this figure it still serves as a useful reckoning point. The savings are calculated in gallons per year to the nearest half gallon.

Overall mile/gall in untuned state	Percentage increase in mile/gall					
	1%	2%	3%	4%	5%	10%
20	$4\frac{1}{2}$	10	$14\frac{1}{2}$	$19\frac{1}{2}$	24	$45\frac{1}{2}$
25	4	8	12	$15\frac{1}{2}$	19	$36\frac{1}{2}$
30	3	6	$9\frac{1}{2}$	$12\frac{1}{2}$	$15\frac{1}{2}$	30
35	3	$5\frac{1}{2}$	$8\frac{1}{2}$	11	14	26
40	$2\frac{1}{2}$	5	$7\frac{1}{2}$	$9\frac{1}{2}$	12	$22\frac{1}{2}$
45	2	$4\frac{1}{2}$	$6\frac{1}{2}$	$8\frac{1}{2}$	$10\frac{1}{2}$	20

savings in gallons per year

Let's assume that between services a car drifts from maximum fuel consumption to a point where its fuel consumption is 10 per cent below par, for example, a car capable of 45 mile/gall will be running at around 40.5 mile/gall. Its average tune will be about 5 per cent below par, representing 42.75 mile/gall. By proper adjustment it is possible to keep the car much nearer its optimum performance, running at an average of say 44.5 mile/gall and from the table above this can be seen to mean a saving of about 8 gallons a year. This might appear to be slim pickings but remember that this amount of petrol can be gained from a very low expenditure, say 75p to £2 on a contact breaker set. And of course the saving is relative to the price of petrol. Also larger savings will be made the lower the car's mile/gall.

Modification of the engine, gas-flowing, matching of manifolds to ports and so on, will give much larger, more permanent gains. On a typical family saloon which, in good tune, returns 30 mile/gall a 10 per cent improvement will save 30 gallons a year. If the life of the car is considered to be $3\frac{1}{2}$ years in a single ownership the total saving from this tuning operation will amount to 105 gallons. As the average price of petrol over the next $3\frac{1}{2}$ years is likely to be as high as £1 per gallon, that's a saving of £105: the outlay to achieve this saving could be as low as £10 to cover grinding tools, materials and apportioned tools costs. The net saving is £95. Even larger savings can be made on larger engined cars.

1:4 Good workshop practice

Tuning and maintenance work must be carried out with care. Many tuning techniques and processes involve work on safety related components of the car like the brakes and suspension. Bad workmanship on these systems is potentially dangerous and, additionally, in being unfit for the road the vehicle may contravene the law. The law is quite specific in nominating the driver as being solely responsible for a dangerous car.

Some tuning operations involve making changes to what the manufacturer sees as the desirable virtues of a vehicle (those that increase the vehicle's reliability and longevity). Tuning can enhance these characteristics of a car; on the other hand, poor workmanship can cause extensive and expensive long term damage.

Rules are tedious but very necessary in tuning a car. Observe the following guidelines below and you could avoid expensive and disastrous mistakes.

1 Never increase the performance of a worn engine without carrying out a thorough inspection and overhaul, including the replacement of stressed parts such as crankshaft bearings, big and little ends, pistons rings and bores, camshaft and bearings, valve guides, rocker arms, pushrods and cam followers. These are the principle parts that must be in good condition. Other weak points to watch are the timing chain and sprockets and the clutch which may not be capable of taking the additional load. Although engine wear is for the most part determined by the number of miles covered by the car since new and between major overhauls it is also dependent on the kind of use the car is put to. Cars used mainly for short town journeys will warrant a full inspection and overhaul at around 20,000 miles before using performance parts. Cars used for a normal mix of long distance motoring and short journeys (provided they are treated respectfully) will probably be in good enough condition to withstand performance treatment at up to 30,000 miles. Hard use of a car in standard condition will accelerate wear tremendously and lower the mileage at which a complete overhaul becomes necessary.

2 Never attempt tuning or performance work on an engine without a workshop manual containing a complete specification of adjustments, tolerances and tightening torques. Most manufacturers keep workshop manuals reasonably up to date with additional sheets for modifications on models. However, check that the manual does contain all the information for the car being tuned. There are in-service modifications to some cars that never appear in the workshop manual – a main agent may be able to supply information on those that could affect tuning work.

3 Because there are continual modifications made to cars without any outward model change taking place, take every step possible to ensure that new and used parts fitted to a car are suitable for use on that vehicle. A minute comparison with the old part is usually the best method.

4 Always work in the cleanest possible conditions. Prior to removal of parts clean down the engine compartment, suspension and transmission with a high pressure hose, remove grease by pre-treatment with Gunk or a neat detergent like Teepol. To protect electrical components like the coil, distributor, control box, generator and starter motor from the water, cover them in plastic bags secured with rubber bands.

Parts can then be removed for further cleaning on the bench. Dismantling of the part should not be commenced until the exterior is free from grease and accumulated mud and dust and the bench itself has been cleaned down.

5 The exploded diagrams in a workshop manual are never quite good enough for correct reassembly of parts. Don't rely on the diagram and memory – where possible mark parts on stripping to make correct assembly easier. Make drawings of assemblies taking particular note of the order of parts like spacers, washers, shims and other small components that might seriously affect the tolerances of the unit. Label electrical wiring and connections to ensure they are reconnected in the right way.

6 When stripping down parts that are going to be reused ensure that those parts which have worn together by mechanical contact are set aside neatly so they can be replaced in their original positions. This instruction particularly applies to the components of the engine. Pushrod, rocker arm and valve parts must always be kept together in their original order as they become matched to one another by wear – reassembling unmatched parts can cause serious damage. The same care should be taken with the piston, connecting rod and crankshaft assembly.

7 When making tuning adjustments never make changes to more than one element at a time – otherwise it becomes impossible to tell which adjustment is affecting the engine's performance.

8 When drilling or removing metal from engine parts ensure that turnings and swarf cannot fall into engine apertures. Wash waste metal out of manifolds and cylinder head ports with paraffin or petrol before reassembly. Cotton wool, tissue papers or clean rags can be stuffed into engine apertures to prevent the ingress of dirt, drill swarf and small parts while it is stripped down.

9 When lifting a car or the engine use a properly designed jack. For maintenance and tuning work never use the original equipment jack. Never leave the car supported on

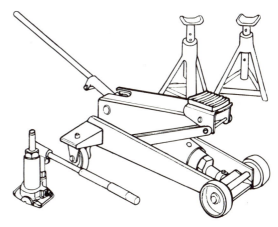

FIG 1:1 Bottle jack, trolley jack and axle stands

the jack – take the strain with axle stands. It is wise to disconnect the battery when working on the car for long periods as shortcircuits can easily occur when components are disconnected – also the starter motor may be inadvertently operated.

1:5 Tools, instruments and their care

Most of the tools used for tuning are those that should be found in any good home mechanic's tool box. All tools should be of good quality – never buy cheap tools, they wear or break perhaps damaging the car or causing injury.

A good tuning tool kit should contain the following items:

Spanners and sockets:

Maintenance and tuning is made much easier with sockets or ring spanners. These cause less damage to the hexagon heads of nuts and bolts and can be used to apply more torque in safety. However, open-ended spanners are needed for nuts in some locations. Buying duplicate sizes in each set may appear wasteful but it is often necessary to use two types of spanner of the same size when tightening unsecured bolts or when adjusting locknuts. Useful size ranges are Metric 9 to 17 mm and AF $\frac{3}{8}$ to $\frac{11}{16}$ inch. A brake adjuster suited to the particular car is the most common special spanner the tuner will require. Socket, ring and open-ended types are available, specially designed to fit the square adjuster head on the wheel backplate.

FIG 1:2   Brake adjusting spanner

Electrical work may require a set of small spanners (sometimes called magneto spanners) and for some cars there are special spanners and wrenches available for removing sump, transmission and engine block oil and water drain plugs.

Plug spanner:

Plugs are best removed using a specially designed plug spanner with a short-handled lever and a rubber insert to prevent damage to the plug insulator. Plug sockets used in conjunction with the socket set wrenches are not a very good idea as, by using long levers, it is very easy to overtighten the plug (on alloy heads this may strip out the plug threads and it can ruin the seal on cast iron heads).

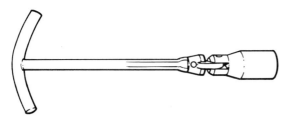

FIG 1:3   Plug spanner

Avoid using an ordinary box spanner, it can cause insulator damage. Check the car's handbook to find the size required – 10 mm, 14 mm or 18 mm.

Allen keys:

A few cars have Allen bolts instead of the normal hexagon headed type especially on key components like the inlet and exhaust manifolds. These bolts are removed with Allen keys usually sold in sets of $\frac{1}{16}$ to $\frac{3}{8}$ inch AF and 1.25 to 10 mm Metric. Some manufacturers (Ford, Opel) use an internal spline for some bolt heads: these will require suitable special tools.

Torque wrench:

The force applied in tightening a nut or bolt is measured in units of torque. The Imperial units are lb/ft and the Metric units are kgm (kilogram metres) or Nm (Newton metres). Practically all nuts and bolts on the car have a specified tightening torque especially those on stressed parts of the engine or gearbox. The torque wrench is an instrument designed to tighten nuts and bolts to the correct torque and it is usually made for use with $\frac{1}{2}$ inch square drive socket set.

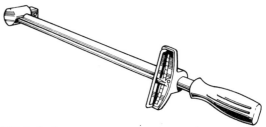

FIG 1:4   Protractor-type torque wrench

The simplest and cheapest type of torque wrench consists of a drive head and a long handle to which is attached a scale. A pointer runs from the drive head to the scale. The application of torque to a bolt bends the handle and the scale moves in relation to the pointer which indicates the force applied. The bolt is tightened until the

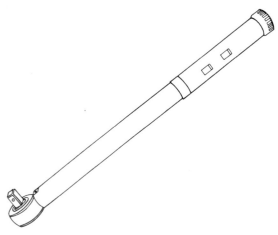

FIG 1:5   Ratchet-type torque wrench

specified torque is reached. This type of wrench is quite difficult to use accurately.

A more expensive type of wrench consists of a ratchet drive with adjustable spring tension. The torque required is preset on a scale (usually by means of a knurled adjuster on the wrench handle) and the bolt is tightened until the wrench is felt to jump with an audible click as the ratchet is over-ridden.

Most of the car's torque setting requirements can be catered for with a wrench having a range of 20 to 110 lb/ft. There are occasional requirements for lower or higher torques – wrenches for these values can be obtained on hire (Yellow Pages under Hire Contractors – Small tools and equipment).

Valve or tappet adjusting tools:

Some cars need a special type of spanner for valve clearance adjustment – check with the car's workshop manual. If the tool is a bent spanner (some Fords and Datsuns) it may be possible to make this tool more cheaply than its retail price.

The valve or tappet clearances on most cars can be adjusted using the SPQR tappet adjuster. This is a simple, almost foolproof combined screwdriver and wrench with a device to measure the travel of the adjusting screw. The chief advantage of using this tool is that the innaccuracy

which can arise in using feeler gauges on worn parts is avoided – the tool takes the wear into account.

Screwdrivers:

A selection of different sizes of cross-head and bladed types of screwdriver is necessary.

Pliers and grips:

Three types of pliers are useful additions to the tool kit. The standard combination pliers have limited use in tuning because of the size of the nose. Finer work, particularly on electrical components needs long-nosed pliers. A special type of pliers is available for removing and replacing circlips.

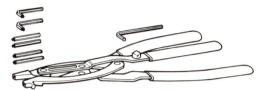

FIG 1:7   Circlip pliers

Feeler gauges:

Use good quality feeler gauges – they come in sets measuring from .001 inch (1 thou) to .025 inch (25 thou). The feeler blades may be marked with the metric equivalent or a special metric set may be purchased with feelers from .02 mm to .60 mm.

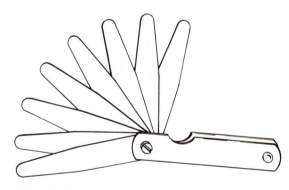

FIG 1:8   Feeler gauges

Spark plug gap tool:

A combined spark plug gap setting tool with .016 inch to .032 inch feelers and a contact file is sold by Champion. It is a worthwhile purchase which makes spark plug setting very much easier.

Micrometer:

For measuring shim thicknesses and wear on shafts a good quality micrometer is essential. Various size ranges are available – most car requirements are covered by two micrometers measuring in the ranges 0 to 1 inch and 1 to 2 inch.

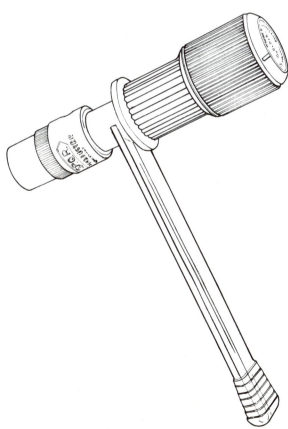

FIG 1:6   SPQR tappet adjuster

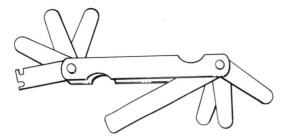

FIG 1:9 Champion spark plug tool

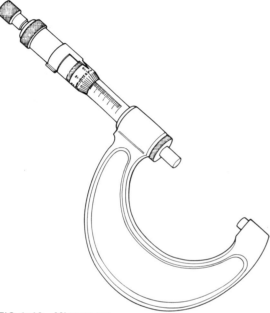

FIG 1:10 Micrometer

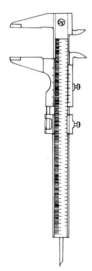

FIG 1:11 Vernier calipers

Vernier calipers:

Internal and external diameters of bearings and shafts require the use of internal and external vernier calipers with a vernier gauge calibrated down to thousandths of an inch.

Stroboscopic timing light:

A timing light enables accurate adjustment of ignition timing according to the timing marks found on the crankshaft pulley or the flywheel/starter gear ring. Timing lights within the price range of the home tuner are of two types. The best has an independent source of current supply for the stroboscopic flashing light (the car's battery) and the pulses are timed by a lead sensing the high voltage impulses to the number 1 spark plug. This kind of strobe has a much brighter flash and makes viewing of the timing marks very easy even in daylight.

A much cheaper type of strobe utilises the spark impulse itself as the power source and the light produced is correspondingly weaker.

FIG 1:12 Stroboscopic timing light

Circuit tester and static timing light:

A good basic circuit tester and static timing light can be made from a 12-volt (or 6-volt if the car has this system) low-powered (5 watts or less) light bulb in a small bayonet or Miniature Edison Screw (MES) lampholder. Connect to one terminal of the lampholder a long lead (4 to 5 ft) with a small crocodile clip at the other end. To the lampholder's other terminal connect a much shorter lead (about 1 ft or less) the other end of which is attached to a sharp pointed probe.

A good probe can be made from a knitting heedle or piece of copper rod up to $\frac{1}{8}$ inch diameter. The probe

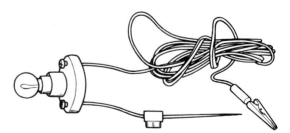

FIG 1:13 Home-made circuit tester

should be about 6 inch long and the point should be sharp enough to pierce the plastics insulation of a wire without too much force. An easy way to connect the probe to the lead is to use a plastics insulated barrel connector. The probe should be insulated leaving only the pointed tip bare. Plastics insulating tape can be used for this although a neater, more permanent, job is made by shrinking thin plastics tube onto the probe – detergent and water will make the tube easier to slip onto the probe. The rest of the connections on the circuit tester should also be well insulated to guard against the possibility of a shortcircuit.

Commercial circuit testers are available which have a neon bulb inside the probe handle.

Tyre tread depth and pressure gauges:

Tyre condition is as important for petrol economy as it is for performance motoring – a reliable tyre pressure gauge is essential and the only good way to measure tread depth is to use a specially designed gauge. A set of the two instruments can be purchased quite cheaply.

Compression tester:

Various types of compression tester are available to the home tuner. The cheapest is similar to a tyre pressure gauge, having a shank that screws into the spark plug hole. Dial types are more expensive but are usually more accurate. Some have a screw adaptor for the spark plug hole, others have a rubber gasket which is pressed tightly over the hole to make a pressure seal.

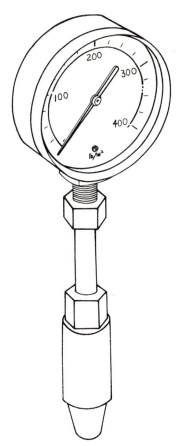

FIG 1:15   Compression tester

Power tools and accessories:

Cylinder head modification and work on manifolds cannot be attempted without a good quality two-speed electric drill. Essential accessories are a flexible drive shaft and sets of grinders of various shapes. Sets containing all the types normally required for this work are widely available at tool suppliers. Not so widely available are the small backing discs necessary to retain small pieces of emery paper for polishing operations – these may have to be made by the tuner.

Tachometer (rev counter):

Many high-performance saloons are already fitted with a tachometer. If this is not a part of the car's standard equipment it is a very useful accessory instrument to fit. For accurate ignition timing a tachometer is indispensable.

Special tools:

A glance at the manufacturer's workshop manual may discourage the home tuner because of the frequent mentions of special tools for practically every operation on the engine, transmission and suspension. In practice there are very few tasks that cannot be performed with quite ordinary tools and a little bit of ingenuity and patience. This is not to say that discriminating use of

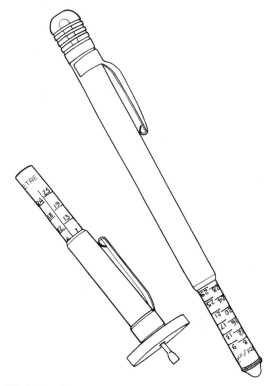

FIG 1:14   Tyre pressure and tread depth gauges

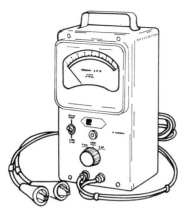

FIG 1:16 Tuning tachometer

special tools cannot make jobs a great deal easier – a few tools are suggested below and in cases where use of a particular aid is unavoidable a tool hire specialist should be consulted.

Hub puller:

To withdraw a wheel hub from over the tightly fitting wheel bearings a puller is necessary. A simple puller is a bar or plate with a centre screw and holes for pulling bolts that align with the holes or studs on the hub – in other words pullers are designed for specific cars. It is sometimes possible to use a universal puller which has adjustable pulling arms – these are expensive items and it is best to hire them from a tool specialist.

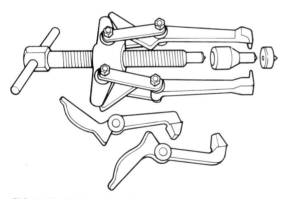

FIG 1:17 Universal puller

Flywheel puller:

Front-wheel drive cars (BLMC models in particular) often require removal of the flywheel to gain access to the clutch. The flywheel is fitted by a tight taper joint to the end of the crankshaft and this joint has to be broken by the use of a considerable pulling force. Flywheel pullers are of a similar design to hub pullers. It is not usually possible to use a universal puller in this application. However, for many makes of car there is a dual purpose hub and flywheel puller available from the manufacturers or the appointed special tool manufacturer.

Pullers of both types can be made up in the home

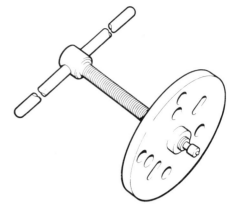

FIG 1:18 BLMC hub and flywheel puller

workshop or by a local engineering works. Home tuners, lacking a proper puller, have used many dodges to overcome the problems. These include roping the part to be removed and exerting the high pressures necessary with a hydraulic car jack. With any such improvised technique take care that nothing can slip during the operation; nasty injuries to the hands can result. Mini owners can make an excellent flywheel puller from the flange of a Hillman Imp drive shaft cut to size. Three high tensile bolts (with nuts already threaded on) inserted in the drive flange holes mate with the threaded holes on the flywheel face – tightening each nut a little at a time slowly draws the flywheel off the taper.

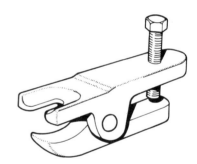

FIG 1:19 Screw ball joint remover

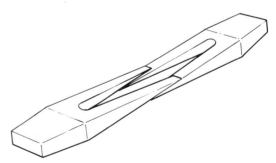

FIG 1:20 Wedge ball joint remover

Suspension and steering ball joint remover:

Ball joints on suspension and steering parts are often fixed by means of a taper jointed pin. Various tools are available from car manufacturers or the appointed special tools supplier to break this joint. A common kind consists of a fixed bar shaped to fit under the ball joint with a hinged lever arm tensioned by a bolt. The hinged arm bears on the end of the pin and tightening of the bolt forces the joint apart. Two opposed impact wedges hammered simultaneously between the ball joint and the steering ·or suspension to which it is fixed is another method widely used.

Special tuning instruments and tools:

Burette:

A 50 ml burette with .1 ml graduations and an appropriate stand will have to be bought or hired for balancing ports and combustion chambers.

Dial gauge:

Precision setting of piston heights and valve timing requires the use of a dial gauge which measures the varying heights of moving components above a zero reference height. Various types are available – the easiest to use have heavy magnetic bases. Other types may have

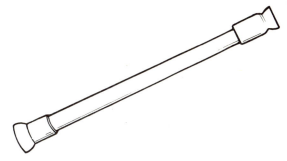

FIG 1:22   Valve suction grinder

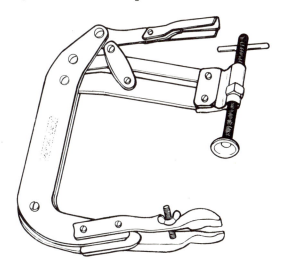

FIG 1:23   Valve spring compressor

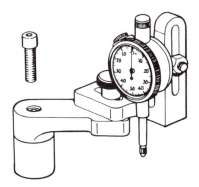

FIG 1:21   Dial gauge

clamping bases to fix the dial over the reference height. This is an expensive instrument and it is best hired from a specialist tuning equipment supplier.

Valve suction grinder:

Lapping in valves is carried out by smearing grinding paste on the valve and seat and then rotating the valve in the seat using a valve suction grinder – a wooden rod surmounted by a rubber sucker which locates on the flat face of the valve. This tool costs a few pence and can be obtained from most car accessory shops.

Valve spring compressor:

Adjustable valve spring compressors make light work of the removal of valves from the cylinder but, most important, they make correct assembly of oil sealing and valve retaining devices possible.

Drifts and punches:

A wide variety of parts secured by interference fit, like bearings and guides, have to be driven in or out of housings and recesses with a great deal of force. Applying the force correctly without causing damage to the part or adjacent areas of the engine is very difficult unless drifts are used which are of a similar diameter to the part. A wide selection of varying diameter steel or brass drifts or punches are a useful addition to the tool box. Sometimes it is possible to make a suitable drift from an appropriately sized piece of bar, tube or wooden dowel.

Care of tools:

The car is probably one of the dirtiest environments in which to carry out any engineering work. Inevitably tools become coated with grease, grime and thick smears of underseal. Unless the dirt is cleaned off after every tuning or maintenance session the filth will be transferred to the tool box. Clean spanners, sockets and other heavy duty tools with a cloth soaked in petrol or turpentine substitute.

Keep more delicate tools and instruments like torque wrenches, tappet adjusters and spark plug tools out of the main tool box preferably in a separate box which can be kept clean and dry.

Torque wrenches with adjustable ratchets should never be left under tension – back off the adjuster to zero or the minimum torque in the range. Delicate instruments like micrometers and vernier calipers must always be kept in a separate case or box and some means of keeping them dry must be used. Special water repellent papers are available from engineering supply specialists – a small piece of this paper in the instrument case will guard against surface rust. An additional protection is to give the instruments a light coat of thin oil.

Examine tools, especially spanners, for wear at regular intervals. Wear on the spanner jaw to the extent that the fit on a nut or bolt is affected means the end of the spanner's useful life. To go on using it could mean the stripping of the bolt head and, very possibily, injury when the jaws slip under high tension.

The heads of hammers must be examined for splitting or chipping. Using a hammer that is in this condition is extremely dangerous – chips can fly into the eye at very high velocity.

Regrind and square off chipped or worn screwdriver blades. Discard worn cross-head screwdrivers – they can do a great deal of damage to screw heads.

CHAPTER 2

Work on the engine

2 : 1 Engine swops
2 : 2 Improving carburation
2 : 3 Cylinder head modification
2 : 4 Camshafts
2 : 5 Lightening, toughening and balancing

2 : 6 The cooling system
2 : 7 Tuning the exhaust system
2 : 8 The lubrication system on performance cars
2 : 9 The ultimate in tuned engines?

In **Chapter 1** it was explained that the manufacturer rarely allows an engine to develop its full power output. The engineering work required to do so would prove far too costly for a mass produced vehicle. Much of tuning is the rectification of this restriction in quality. Measures directed at economy rarely differ from those aimed at increasing performance – the methods are the same but the extent to which they are applied is different. In the following chapter on the parts of the engine which can be modified or replaced the point at which a particular process becomes uneconomic as a result of the cost of modification or the increasing fuel consumption should be noted.

2 : 1 Engine swops

One of the simplest ways to gain extra performance is to look up the manufacturer's model range for an engine that will give the required results. If at the same time the existing engine needs replacement, economies can be made in this process too. By removing an engine that endowed the car with a lower power : weight ratio (generally needing high revs to achieve reasonable speeds through a low differential ratio) and putting in an engine one size larger with its appropriate differential it is usually possible to make a gain in mile/gall. This is always provided that the extra power that becomes available is not used. Even if a gain in fuel economy is not achieved the car will be much pleasanter to drive.

Swapping an engine for power alone increases the power : weight ratio. The limit to the possible increases depend on the engine used and the ingenuity of the operator but it is beyond the scope of this book to describe the massive engineering changes that must be undertaken to slot a Jaguar engine into the bonnet of a

Mini. Wilder swops of this kind have no hard and fast rules – few are ever carried out that result in an easily driven, tractable car suitable for day to day use.

Putting a larger engine into a car increases the work that various ancillary units and systems have to carry out. Here are the principal points where strain arises:

Gearbox:

The bellhousing may be the point at which the new engine is mated to the existing transmission – check that it is of the proper shape. It is often advisable to swop the gearbox with the more powerful engine – the gearbox may have closer ratios and will therefore be a worthwhile addition to the car. Use the clutch size and type that is standard on the larger engine.

Transmission:

Prop shaft length may become a problem if there is any difference in engine and gearbox length between new and old units. The performance engine's prop shaft may have to be used or, if the swop is unusual, involving the use of another manufacturer's engine, a special may have to be made up to the correct length with the appropriate universal joints. It will almost certainly be necessary to make a change in differential ratio to take advantage of the new power available. It may also be possible to consider using slightly larger wheels.

Brakes:

Standard braking systems usually have a reasonable reserve of stopping power that can cope with the higher speeds and more repeated use demanded for a performance car, but more reassured stopping power will be obtained by uprating the system. One solution might be to fit harder brake linings – a servo will be necessary to restore to normal the higher pedal pressures that result from this change (see **Chapter 6**).

Suspension:

There are very few cars that can accept a larger engine without requiring some consideration of the suspension characteristics. The faster a car is able to go the more important it is that the suspension is modified to reduce the task of controlling the vehicle. The minimum requirement in fitting a larger engine is to uprate the shock absorption – fitting an anti-roll bar and uprating the springs are two further moves that should be considered (see **Chapter 5**).

Cooling system:

The higher the power output of the engine the greater the waste heat output that the cooling system has to deal with. Generally, there is a fairly high reserve of heat dispersal in the standard cooling system applied to the smaller engined cars. This may be sufficient to cope with a size larger engine and give a useful rise in average cooling water temperature – however this reduces the margin for error which exists to allow for extra hot days, long hill climbs and so on. Consider fitting the appropriate radiator for the high power engine or having the existing unit rebuilt with a more efficient core by a radiator specialist. The heat problem also extends to the temperature of oil. Recently, specialist opinion has been against the fitting of an oil cooler as an instant panacea for oil problems. Provided that the engine's original sump capacity is preserved, very few problems will arise in normal motoring. If a smaller sump pan has been fitted to clear a crossmember, it may be extended in depth at one end to restore the capacity.

Exhaust systems:

Always use the exhaust manifolding and system from the higher powered engine – if this results in clearance problems during fitting, specialist tuners will almost certainly be able to supply an off-the-shelf manifold which will be suitable.

Ancillary equipment:

Carburetters, inlet manifold, starter motor and distributor must all be transferred with the engine or chosen to suit the new unit. It may be necessary to design special throttle linkages to operate the carburetter of a new unit – this is never as complex a problem as it may at first seem. Carburetter manufacturers may be able to help solve problems but generally a trip to the scrapyard will provide enough ideas and parts to manufacture an excellent mechanism.

Bodywork modification:

If the engine from a different model range is used it is often necessary to make some modifications to the bodywork. These may simply consist of the wielding of a large hammer to create bulkhead bulges which will accommodate a larger bellhousing. A common necessity is for a repositioned gearlever hole – remember to blank off the old hole well, as a great deal of noise can enter the passenger compartment in this way. It is also frequently necessary to change crossmembers under the engine or at the rear gearbox mounting – suitable parts can often be found on other models in the range or slight modification of the existing part can be made. If parts have to be cut away make suitably amended replacements and weld them back in place – this is particularly important in the case of sections cut from the engine compartment bulkhead. This area contributes substantially to the strength of unitary construction car bodies and is a vital first line of defence against the engine's entry into the passenger compartment in an accident.

Speedometer:

A change in engine, gearbox, differential or wheel size will affect the calibration of the speedometer. Usually the answer is to swop the speedometer head with the power unit; however changes in the speedometer drive gear can be made – any persistent calibration problems should be referred to the instrument manufacturer who will be able to advise on the best practical solution.

Suggested engine swops:

Fords:

Putting a 1300 engine into an 1100 Escort is a useful economy modification. For performance the 1300GT engine can be used – there are no problems at all with this modification. Using the 1600 and 1600GT engines from the Cortina involves making changes to the sump (or an Escort Mexico sump can be used). Gearboxes become a problem – a suitable choice is the strong unit from the Corsair (V4 model) or the Cortina unit. An Escort GT exhaust system has to be used as the Cortina's will not fit.

It is also possible to make similar swops on the Cortina range, going up one model engine size. Owing to the greater diversity of engine types in this range, there are many more complex problems in designing or finding engine mountings and ensuring crossmember clearance. Remember that there was a change to cross-flow engines on the Mk 2 Cortina and that this introduces problems of arranging the exhaust and inlet manifolding and the carburetter linkages.

Vauxhall:

Older models of Viva in the HB and HC ranges will accept the later 1256 cc engine with no problems whatsoever. It is not, however, possible to fit the larger units that have been used with the car – 1600 to 2300 cc units require changes throughout the transmission.

BLMC:

Small Triumphs, the Herald and Spitfire, will accept the later 1300 Spitfire twin carburetter engine – changes to engine mountings and exhaust system (use a specialist manifold for the Herald) are the only problems. Using the 1500 engine may necessitate a gearbox change. Both cars will accept the engine and gearbox of the 2000, GT6 or Vitesse provided the rest of the transmission, particularly the rear axle, is changed too.

1.3 Marinas will accept the 1.8 engine and gearbox with a potential gain in economy.

Morris Minor 1000 models will accept an 1100 cc A series engine but for real performance a 1275 cc engine from a later Midget or Sprite will be necessary – use the sports car gearbox too. A40s will accept the same engine.

Perhaps the greatest number of variations of any car can be played on the Mini. 1100, 1275 and 1300 engines can all be fitted to the 850 or 1000 gearbox provided that the gearbox casing is ground away to allow sufficient clearance for the larger crankshaft. This can be performed quite simply by liberally coating the crankshaft web ends with grease, offering the engine to the gearbox and rotating the engine by hand. Remove the engine and scribe the grease marks on the gearbox casing. Grind away the areas marked sufficient to allow adequate clearance. Ensure no grindings enter the gearbox by using a vacuum cleaner with the nozzle very close to the grinding area – also stuff cloth into all the spaces possible. Clean down the whole box with a liberal dose of petrol after the work is completed. A 1275 engine will now bolt straight on – the front engine plate of the original engine will have to be transferred to an 1100 or 1300 engine so that the engine mounting will fit and it will be necessary to create a clearance for the crankshaft damper. Install the 3.44 : 1 differential of the Mini Cooper as well.

Swopping engines and gearbox is more complicated because of the variations in mountings, drive shafts and differentials. The 1275 Cooper S engine will not fit straight in without changes to the entire transmission, hub, brake and wheel systems. All larger engines will require a more efficient radiator – the Cooper type will do.

A simpler change can be made using the 1100 or 1300 engine. Some modification of the remote control gearchange mountings will have to be made and engine mounting plates will have to be changed for those of the Mini. The differential must again be changed.

2:2 Improving carburation

One of the most popular ways of **1**, achieving better filling of the combustion chambers with petrol/air mixture and, **2**, obtaining a more even control of the mixture, is to fit a larger carburetter, a precision-engineered unit or multiple carburetters. The ideal situation for performance is to have a carburetter per cylinder.

On specialised competition cars and on some production cars the ideal is achieved by the use of an eight-port cylinder head (for a four-cylinder engine). This design allows the fitting of, for instance, a bank of four Amal carburetters, compact units each one supplying a separate cylinder. A much more common design of cylinder head has siamesed ports in which a single port serves two cylinders in which case two carburetters (three on a six-cylinder engine) can be fitted. In many cases a satisfactory increase in performance – or improved economy – can be obtained by fitting two units of the same type as that fitted singly to the standard car. There are designs of carburetter, widely favoured among performance tuners, that are in effect two carburetters in one.

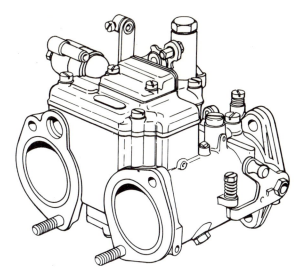

FIG 2:1   Dellorto twin choke carburetter

The well-known Weber and Dellorto twin choke models are of this type.

The most modest performance and potential economy tuning measure is simply to fit a larger version of the standard carburetter.

In most cases the fitting of a larger carburetter or twin units to a car will require the purchase of a matched inlet manifold design (see later section). Other problems include the provision of suitable linkages to operate the carburetter.

The following sections cover a wide range of carburetters that are fitted as standard to some cars and are potential power or performance modifications to many vehicles.

Where possible the advantages to be gained by fitting the units, tuning procedures and hints are given together with a brief description of the operating principle. For tuning details of standard carburetters refer to the relevant workshop manual – accessory carburetter manufacturers will supply detailed tuning data which in most cases covers jet modification required for a selection of special combinations of compression ratio and other tuning modifications. However, on some advanced carburetter designs, like the Weber DCOE types, expert assessment of the car's performance characteristics is required to determine the correct values for various carburetter parts. Only tuning shops expert in the particular carburetter will be able to do this. A book like this cannot possibly cover all the possible combinations of cars, modifications and carburetters.

SU carburetters:

Most BLMC cars are fitted with SU carburetters – British Leyland own the SU Carburetter Company. The SU design is of the type known as variable jet in which the size of the choke (air passage) and the petrol metering jet are both varied automatically.

The SU carburetter has a float chamber with a float valve to maintain the pressure of the fuel supplied to the metering jet at a constant pressure equal to atmospheric.

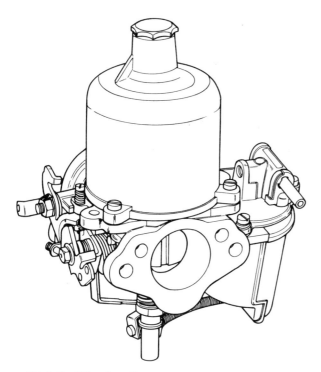

FIG 2:2   SU carburetter

A flexible link tube connects the chamber with the jet body – the jet orifice size is controlled by a tapered needle which slides up and down inside the jet body.

Movement of the needle is effected by the movement of a piston in a chamber evacuated by the manifold depression. As the engine demands more fuel, the piston is sucked up inside the dashpot drawing the needle out of the jet orifice and effectively increasing the jet size. Thus more petrol can flow out. Airflow through the carburetter is controlled by a butterfly (see **FIG 2:3**).

It is the taper of the needle – manufactured to extremely fine tolerances – that determines the performance characteristics of a car fitted with an SU. Cars with the unit as standard are fitted with a needle that gives an overall compromise between performance and economy. Workshop manuals will state this as the standard needle – optional weak running and rich running needles will also be given.

Latest SU carburetters have non-interchangeable needles – in keeping with other anti-emission measures SU have designed all current needles to give the best performance consistent with low pollution levels in the exhaust. These spring biased needles have the advantage that they are slightly weaker than the previous standard needle designs.

For older SU equipped cars the first economy measure is to fit a weak needle. This is a simple job – the needle securing screw in the piston is undone and the new needle slipped in place. It is fixed with the shoulder flush to the piston and centred as described in the next paragraph. This is a reasonably effective measure; the only disadvantage is that the greatest economy is made at higher speeds due to the taper design.

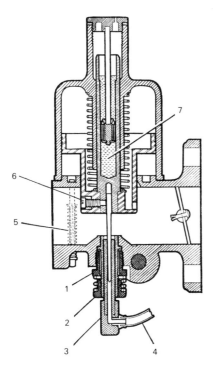

FIG 2:3   Cross-section of SU carburetter

Key to Fig 2:3   1 Jet locking nut   2 Jet adjusting nut
3 Jet head   4 Nylon feed pipe   5 Piston lifting pin
6 Needle screw   7 Piston damper oil well

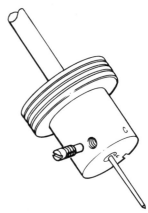

FIG 2:4   Fitting an SU needle

Tuning pointers for SU carburetters:

1 After fitting a new needle the jet has to be properly centred. Undo the jet linkage, remove the jet and take off the adjuster nut and spring – loosen the jet locking nut. Remove the damper piston from the dashpot. With a thin dowel or pencil inserted in the dashpot push the piston down as far as it will go – insert the jet and push that as far into the carburetter as it will go. Tighten the jet locking nut (see **FIG 2:6**).

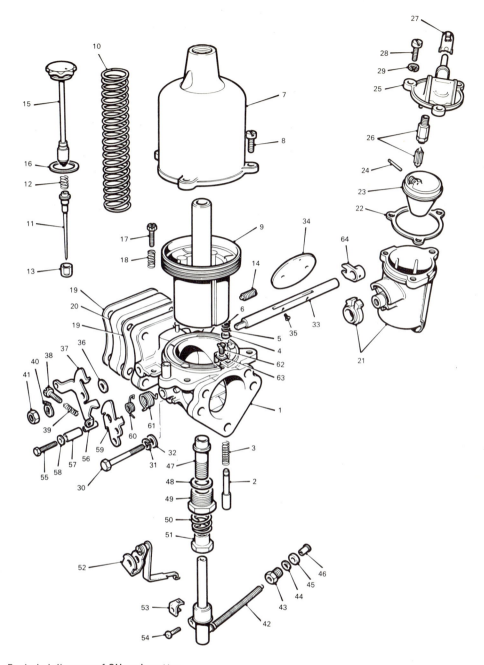

FIG 2:5 Exploded diagram of SU carburetter

Key to Fig 2:5   1 Body   2 Piston lifting pin   3 Spring for pin   4 Sealing washer   5 Plain washer   6 Circlip
7 Piston chamber   8 Piston chamber screw   9 Piston   10 Spring   11 Needle   12 Needle spring   13 Needle support
guide   14 Locking screw, needle support guide   15 Piston damper   16 Sealing washer, damper   17 Throttle adjusting
screw   18 Spring for screw   19 Joint washers   20 Insulator block   21 Float chamber and spacer   22 Joint washer,
chamber   23 Float   24 Hinge pin, float   25 Lid, float chamber   26 Needle and seat   27 Baffle plate   28 Screw,
float chamber lid   29 Spring washer   30 Bolt, securing float chamber   31 Spring washer   32 Plain washer   33 Throttle
spindle   34 Throttle disc assembly   35 Screw, securing disc assembly   36 Washer, throttle spindle   37 Throttle return lever
38 Fast-idle screw   39 Spring   40 Lockwasher, throttle spindle nut   41 Nut, throttle spindle   42 Jet assembly   43 Sleeve
nut, jet flexible pipe   44 Washer   45 Gland   46 Ferrule   47 Jet bearing   48 Sealing washer   49 Jet locating nut
50 Spring   51 Jet adjusting nut   52 Pick-up lever   53 Link, pick-up lever   54 Screw, securing lever to jet   55 Pivot bolt
56 Pivot bolt tube, inner   57 Pivot bolt tube, outer   58 Distance washer   59 Cam lever   60 Spring, cam lever   61 Spring,
pick-up lever   62 Guide, suction chamber piston   63 Screw, securing guide   64 Lost motion lever

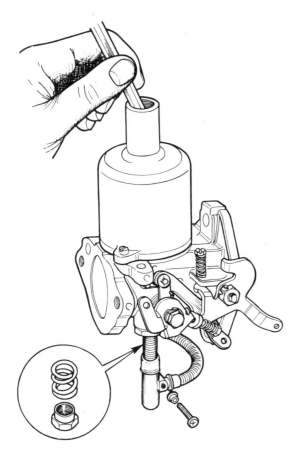

FIG 2:6   Centring the SU jet

Test the centring using the lifting pin to raise the piston – it should fall quickly with a sharp click. Repeat the centring if no click is heard. A few carburetters have no lifting pin – the piston can be raised using a screwdriver.

Refit the parts to the carburetter.

2 Top up the dashpot with SAE20 oil.

3 Mixture setting is carried out by running the engine to normal operating temperature – the idling speed should be about 500 to 750 rev/min. Screw the jet adjuster nut right up and then back down about two whole turns. After the engine has settled to idle, screw the nut up and down until an even, fast idle is obtained. Check the setting by using the lifting pin to raise the piston by about $\frac{1}{32}$ inch. If engine speed rises, the mixture is too rich – if it falls, the mixture is too weak. Find an adjuster nut position between these two at which lifting the piston causes a momentary rise in engine speed followed by a return to normal running. The air cleaner should be in place for these operations – if there is no lifting pin the best idea is to use a carburetter tuning aid. The Colortune described in a later section would be the most suitable.

4 For economy tuning, particular attention should be paid to the operation of the float chamber inlet valve which is effected by the operation of the float. Check the wear on this valve and on the float lever by inverting the lid of the chamber with float attached – the clearance between the float and lid should be $\frac{1}{8}$ inch. A drill shank of the correct size is a handy feeler (see **FIG 2:7**).

The SU carburetter is a very reliable unit that is not prone to wear problems. However, at 25,000 to 30,000 miles it is wise to carry out an inspection strip down examining parts for damage and wear. Particularly vulnerable are the butterfly spindle bearings – replace needle, jet, float chamber valve and float chamber gasket. Lubricate all linkages and pivots with light oil.

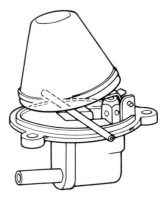

FIG 2:7   Checking SU float level

Tuning the SU for economy:

As described above the economy needle is the best single tuning modification to be made to the SU for the sake of economy. The next best idea is to use a larger SU in place of the standard unit or to install twin SU carburetters. The only objections to this move on economy grounds are that the extra power is bound to be used at times and that the modification could be costly unless it is carried out with good quality scrap parts.

Tuning the SU for performance:

A larger SU and inlet manifold or twin SUs are the best performance measure to take – little will be gained by tweaking the existing carburetter with a richer needle. Parts to carry out both modifications are widely available on higher performance versions of models in the BLMC range – the carburetters will have to be correctly needled for the new application. The SU company will provide technical advice on the needling to be used for various tuning set-ups. Some further tweaking of this robust functional design is possible. For instance the needle is not the only adjustable factor in the carburetter – the spring which controls the rate of lift of the piston can also be

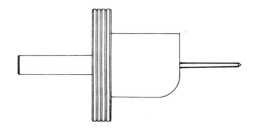

FIG 2:8   Modified SU piston

changed. Once again the manufacturers will help with applications advice.

While the carburetter is dismantled for servicing there is one very quick modification that can be made – radiusing the leading or air inlet edge of the piston. Most older pistons can be radiused over just under half the circumference with a maximum radius of $\frac{3}{8}$ inch at the choke tube centre line (see **FIG 2:8**). Files and wet-or-dry paper are the only tools necessary for this job.

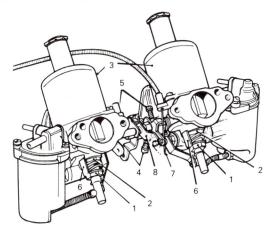

FIG 2:9 Twin SU carburetters

Key to Fig 2:9 1 Jet adjusting nuts   2 Jet locking nuts
3 Piston suction chambers   4 Fast-idle adjusting screws
5 Throttle adjusting screws   6 Jet adjustment restrictors
7 Choke cable   8 Jet lever

Tuning twin SU carburetters:

Very often the single most complex and difficult to perform operation in twin carburetter tuning is the adjustment of the linkages to ensure perfect synchronisation. In this there is no substitute for regular checking of wear and play in the linkages and reference to the individual procedures detailed in the car's workshop manual. General tuning points are as follows:

FIG 2:10 Jet setter device

1 Remove the dashpots and pistons and tighten up the jet adjuster until the jet is flush with the choke bridge – check with a straight edge. Wind down the adjuster the specified amount on each carburetter; alternatively use a device like the Jetsetter to ensure that the jets are adjusted exactly the same.

2 Run the car and, using a device like the Carbalancer or a home-made stethoscope (a piece of $\frac{1}{4}$ to $\frac{3}{8}$ inch plastics tubing), measure or listen to the sound of the airflow in the carburetter. The readings of airflow or the sound has to be exactly the same at each air inlet.

3 Check for correct mixture using the same piston lift test as for a single SU, or a Colortune device (see later).

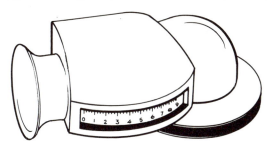

FIG 2:11 Carburetter balancing device

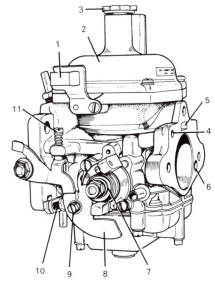

FIG 2:12 Stromberg CD carburetter

Key to Fig 2:12   1 Choke cable clamp   2 Vacuum chamber cover   3 Damper cap   4 Float chamber vent
5 Air vent under diaphragm   6 Temperature compensator vent   7 Choke   8 Fast-idle cam   9 Choke cable connector   10 Fast-idle screw   11 Throttle stop screw

Stromberg CD carburetters:

Strombergs are made and distributed by Zenith and are fitted to Triumphs, Chrysler's Imp Sport and some Volvos among many other makes. The design is on the same principles as the SU variable jet type. The major difference is that the Stromberg uses a rubber diaphragm instead of the piston used on the SU (see **FIG 2:13**).

Fuel enters the instrument at the float chamber which has double floats on a pivot yoke. The yoke impinges on the inlet valve in the same way as the SU. The float chamber forms an integral part of the carburetter base, so the main jet, rising centrally from the chamber, receives fuel direct. The fuel level is the same as that of the jet orifice.

The fuel metering is performed by a tapering needle made to the same high standards as the SU device. It is seated in the air valve (this runs like a piston in a guide tube) which is centrally located in a flexible rubber diaphragm. The diaphragm is clamped in the carburetter housing by the top cover. Once again there is an oil filled damper and a return spring for the air valve/diaphragm assembly.

Cold starting is arranged slightly differently on the Stromberg than on the SU. On the standard CD (the initials stand for constant depression) the choke control operates a lifter bar which pushes up the air valve sufficient to enlarge the jet orifice to give a richer starting mixture (see **FIG 2 : 14**).

CDS types, as fitted to the 1725 cc Hillman Hunter and Triumph GT6, for example, have a different enrichment device. It consists of a disc drilled with orifices which conducts fuel from a separate float chamber feed. There are four varying diameter outlet holes which, according to the choke control position feed varying amounts of fuel into the throttle body (see **FIG 2 : 15**).

The CDSE type is a much later development of the CDS, designed to provide emissions control to meet various national legislation on exhaust pollution. There is no mixture control as such on this carburetter. An idle trimming screw (see **FIG 2 : 16**) provides sufficient adjustment on an air bleed to allow minor mixture

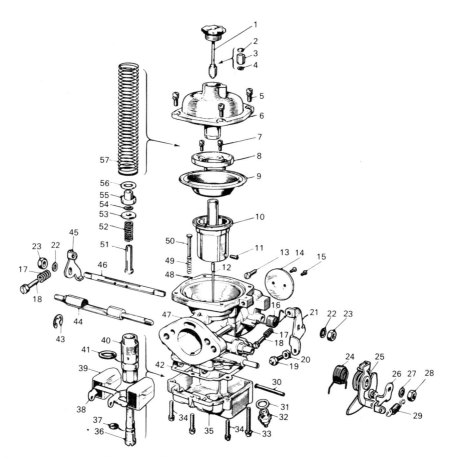

FIG 2 : 13   Exploded diagram of Stromberg CD carburetter

Key to Fig 2 : 13   1 Hydraulic damper   2 Washer   3 Bush   4 Retaining ring   5 Cover screws   6 Suction chamber cover   7 Retaining ring screws   8 Diaphragm retaining ring   9 Diaphragm   10 Air valve and guide   11 Metering needle locking screw   12 Metering needle   13 Choke cable clamp screw   14 Throttle flap   15 Throttle flap screws   16 Throttle return spring   17 Throttle stop screw spring   18 Throttle stop screw   19 Fast-idle control screw   20 Locknut   21 Fast-idle and throttle stops   22 Lockwasher   23 Throttle spindle nut   24 Starter bar spring   25 Choke cam lever   26 Choke lever   27 Lockwasher   28 Choke lever nut   29 Choke lever spring   30 Float fulcrum pin   31 Needle seating washer   32 Needle seating   33 Float chamber screw, long   34 Float chamber screws, short   35 Float chamber   36 Jet adjuster   37 Sealing ring   38 Float arm   39 Float   40 Jet bush retainer   41 Sealing ring   42 Gasket   43 Starter bar retainer   44 Starter bar   45 Throttle stop   46 Throttle spindle   47 Main body   48 Air valve lifting pin clip   49 Spring   50 Air valve lifting pin   51 Jet   52 Spring   53 Washer   54 Sealing ring   55 Bush   56 Washer   57 Air valve return spring

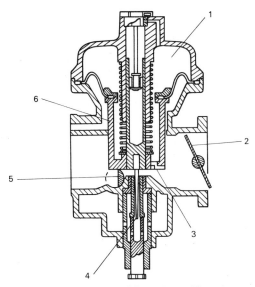

FIG 2:14   Cross-section of Stromberg CD carburetter

Key to Fig 2:14   1 Vacuum chamber   2 Throttle valve   3 Depression transfer hole   4 Needle   5 Air valve lifter bar   6 Air valve piston

strength change to permit compensation for engine age in conjunction with a carbon monoxide meter. A temperature compensation device exerts automatic control by a bi-metallic strip and tapered plug in an air bleed – this adjusts the mixture finely to allow for temperature change of the carburetter body.

The code numbers of Strombergs, 125, 150 and 175, signify choke tube sizes of $1\frac{1}{4}$ inch, $1\frac{1}{2}$ inch and $1\frac{3}{4}$ inch respectively. There is a further vital difference between these carburetters. Owing to the need to vent the air valve diaphragm interface there is a drilling through from the atmosphere flange – an air filter must be selected which takes into account the need to keep this hole open and the varying position of the vent on each size of carburetter must also be noted.

Tuning points for Stromberg CD and CDS:

1   To set the idling speed and mixture strength, remove the air cleaner and take out the damper rod. Use a pencil or dowel to hold the air valve down (see **FIG 2:17**). Screw up the jet adjuster using a coin on the slotted head until the jet is felt to contact the air valve then screw it down three full turns. Run the engine to normal working temperature and set the idle speed with the throttle stop screw to 600 to 650 rev/min. Then carefully adjust the jet screw up and down to achieve a position at which the engine idles with an even beat. Check the setting in the same way as for the SU using a thin screwdriver to lift the air valve for no more than $\frac{1}{32}$ inch. Since there could be a slight enriching of the mixture once the air cleaner is refitted it is best to carry out the setting with a Colortune device.

2   Like the SU, the Stromberg is a reliable unit, but points to watch are the condition of the diaphragm – always look at it against the light to check for minute perforations. Clean the air valve thoroughly as it can stick in the guide.

3   Needles are easily replaced like on the SU and once again jet centring must be carried out. Fully lift the air valve, tighten the jet assembly and then screw up the adjuster so the orifice is just above the bridge level in the choke tube. Slacken the jet assembly slightly and allow the air valve to fall – this process actually centres the jet. Retighten the jet assembly and check the air valve still falls freely. Carry out full adjustment procedure as in **1**.

4   Filling the damper is a little more complicated than on the SU as the oil has to be placed in the tube which is mounted in the air valve. The best way to do this is to hold the air valve up while filling. SAE 20 oil is the ideal damper fluid.

5   Float level should be measured with the carburetter inverted. The proper setting should be 17 mm measured from the gasket flange (gasket fitted) to the base of the float. Float adjustment can be carried out in two ways – bending the tag which contacts the valve or fitting a washer on the valve seat.

6   Adjusting twin Strombergs should be carried out in the same way as for twin SUs.

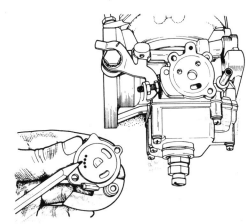

FIG 2:15   Stromberg orifice disc starting device

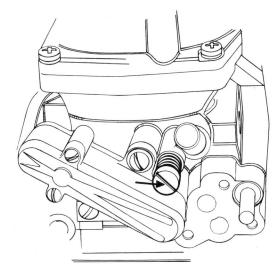

FIG 2:16   Stromberg CDSE air bleed screw

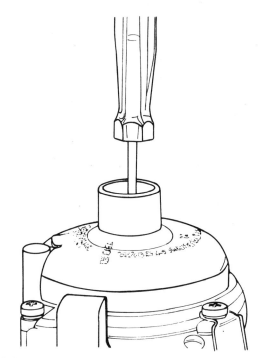

FIG 2:17 Holding down the air valve

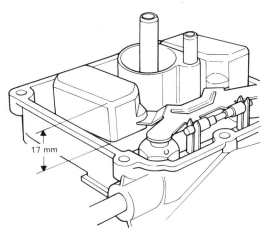

FIG 2:18 Checking Stromberg float height

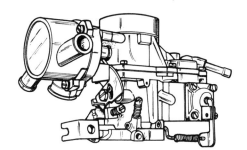

FIG 2:19 Motorcraft GPD carburetter

Tuning Strombergs for economy:

As with the SU, the cheapest and easiest economy measure on a Stromberg is to fit a weaker needle – the car's main dealer or a carburetter specialist will determine the maximum possible increase in needle size.

Tuning Strombergs for performance:

Richer needling, or a taper chosen for certain characteristics may aid performance in some respects but the greatest gains can be made by fitting twin carburetters. Suitable manifolding will have to be used.

Autolite, FoMoCo, Motorcraft carburetters:

Recent Fords have been fitted with a Ford designed carburetter which has been branded with all the above names and is officially known as the Ford GPD downdraught type. It is an unsophisticated fixed choke carburetter, mass produced by semi-automated methods which have generally reduced the amount of work that the home tuner can carry out.

The carburetter has two main castings – the upper and lower bodies. In the upper body are the float pivots, needle valve, main and power valve systems, the idling system and accelerator pump discharge nozzle and it incorporates the air intake.

Some models have an automatic choke mechanism attached to the upper body. The lower body contains the float chamber choke barrel, idling discharge orifice and progression slot, adjuster screws, accelerator pump and distributor vacuum connection.

The very small drillings used in this design frequently cause dirt clogging faults. The carburetter can be dismantled for cleaning purposes but very few of the components that suffer from this fault are replaceable. Removable jets are not generally fitted.

The carburetter has an integral economy device which is a diaphragm operated by vacuum – high vacuum at cruising speeds causes a valve to weaken the fuel supply in the main valve system. Low vacuum on a demand for engine power allows a boosted main fuel supply to flow – thus heavy throttle use should be avoided in the interests of fuel economy.

FoMoCo tuning pointers:

There are a number of model differences on various carburetters in the GPD range, but the common factor is that there are only two adjustments:

1 Warm up the engine to normal operating temperature. Screw the volume screw fully in and then back out two turns. Set the idle at about 750 rev/min. Adjust the volume screw slowly in each direction until increased speed and smoother running is sensed – screw the adjuster as far out as possible without upsetting the engine's running to get good economy. Once the volume screw is set, restore the idling speed to normal.

2 An important check from an economy point of view is the float level setting. With the upper body upside down, check that the housing rim to float base gap is 28.7 mm. The tag on the pivot can be bent to achieve this. Repeat the measurement with the upper body the right way up – it should be about 35 mm and can be adjusted with the second tag (see **FIGS 2 : 20** and **2 : 21**).

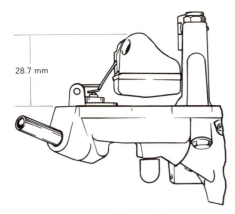

FIG 2:20   GPD float level, inverted

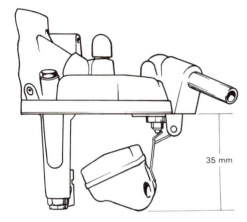

FIG 2:21   GPD float level, right way up

FoMoCo tune for economy:

Adjusted well and maintained in good condition this carburetter is reasonably economical – certainly it would be difficult to recommend a cheap replacement that would do as well bearing in mind the cost of a conversion.

FoMoCo tune for performance:

Models further up the Ford range use Weber carburetters and these would certainly provide an excellent increase in power. There is nothing that can be done to the original carburetter to obtain better performance.

Weber DCD, DFE and DFM carburetters:

Weber have created a reputation for performance carburetters that are second to none. Their main advantage is terrific versatility. As well as cheaper single choke designs used on a wide range of production cars, the company has attracted a great deal of attention from tuners with the twin choke DCD and DCOE ranges.

The downdraught DCD was fitted as standard to a number of Ford high performance models – cheaper versions of the carburetter known as the DFE and DFM types have recently been fitted to some Escorts and Mk 2 Cortinas. DCDs are also fitted to a number of standard Fiat models.

The design is virtually two carburetters in one and both are quite big and excellently designed from the point of view of getting large gulps of air into the combustion chambers. The advantage of having this airflow capability is that the chokes can be arranged to supply the petrol/air mixture progressively. In other words, the first fraction of throttle pedal travel only brings one of the two barrels into operation. When the first choke is two thirds open, the second one comes into play.

The benefits are really much more than this, as the Weber can be fitted with a wide range of choke sizes, and jets. A typical Weber DCD with the widest possible applications is the 28/36 which means that one barrel is 28 mm across and the larger (secondary) barrel is 36 mm across. Venturi and choke sizes are such that the enhanced depression obtainable from the smalle choke allows far more precise control of fuel entering at lower speeds – hence more economy. The disadvantage is that unless a great deal of time is devoted to the study of this device there is very little that the home tuner can do to it. Experience and quite a lot of trial and error are likely to be required to hit upon precisely the right combination of choke and jet sizes for a particular engine and a particular state of tune.

The answer to this problem is that the majority of the tuning shops selling Webers will pre-jet and set up a carburetter to suit a particular application. It will come in a kit form with appropriate manifolding and other fitting items.

The cheaper Ford DFE and DFM types have different cold-start devices and accelerator pump design. The accelerator pump is a device used in fixed jet carburetters to overcome the problem that air accelerates a lot faster than fuel in response to demand. When the accelerator is floored a small pump device ensures that a squirt of fuel is blown into the carburetter to overcome the flat spot resulting from the lull in the supply of fuel. The Weber 28/36 has a piston type pump; the Ford fittings have a small diaphragm unit. In the Ford types the chokes are not removable.

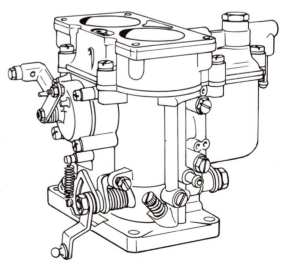

FIG 2:22   Weber 28/36 DCD carburetter

Tuning pointers for Weber DCD:

1 Adjust the throttle stop screw until the engine is at a fast tickover (about 900 to 1000 rev/min) and running at normal temperature. The idling mixture adjusting screw (volume control) is at the base of the carburetter – turn it clockwise until the engine is hunting. Note the amount of turn between these two conditions and try and achieve a midway point at which the engine runs extremely smoothly. A Colortune device will almost certainly make this task easier. When a satisfactory result has been obtained restore to 750 rev/min idle.

2 There may be two volume controls fitted to certain carburetters of this type in which case the only way to establish the optimum point, other than fuel and performance testing, is use a vacuum gauge. Adjust for maximum depression.

3 Float valve and float level are well worth checking on a carburetter of this type. Look out for ridging on the taper

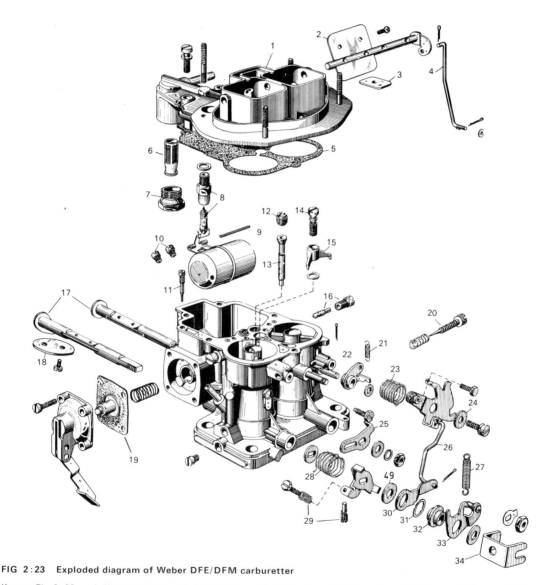

FIG 2:23 Exploded diagram of Weber DFE/DFM carburetter

Key to Fig 2:23   1 Horn assembly   2 Choke plate   3 Dust seal   4 Choke rod   5 Gasket   6 Filter   7 Filter plug   8 Needle valve assembly   9 Float and hinge pin   10 Primary jets   11 Accelerator pump discharge needle   12 Primary air jet   13 Fuel jet   14 Accelerator pump valve   15 Discharge nozzle   16 Idle jet and holder   17 Throttle shafts   18 Throttle valve   19 Accelerator pump diaphragm   20 Idle mixture adjustment screw   21 Choke lever spring   22 Choke lever   23 Choke lever return spring   24 Choke control lever   25 Full throttle stop lever   26 Fast-idle control rod   27 Throttle lever spring   28 Throttle return spring   29 Fast-idle adjustment screws   30 Choke and throttle interconnecting lever   31 Wave washer   32 Throttle shaft bush   33 Throttle lever   34 Throttle link

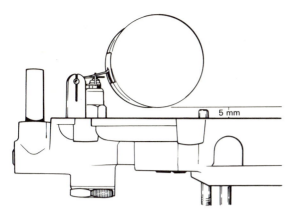

FIG 2:24 DCD float level setting, inverted

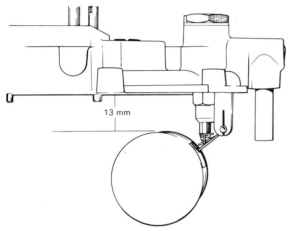

FIG 2:25 DCD float level setting, right way up

of the float needle valve – replace if there are any doubts about its condition. With the carburetter held upside down check the clearance between the top face of the carburetter and the top of the float – it should be 5 mm (DFM = 6.5 mm; DFE = 7.25 mm). If it needs adjusting, lightly twist the needle contact tag. With the carburetter the right way up, the gap should be 13 mm (DFM = 8 mm).

Tuning the DCD for economy:

Used as a kit conversion for an existing carburetter of inferior breathing characteristics it is quite probable that the DCD will record some small improvement in mile/gall – but with all that power there, it will be difficult to resist using it and getting considerably worse figures than standard. In its original equipment format the carburetter is reasonably economical. But carburetter conversions of this kind, bought new, are prohibitively expensive for economy modification.

Tuning the DCD for performance:

Despite the extreme difficulty of buying anything like the right 28/36 DCD set-up for a particular set of modifications first time, this is one of the really effective

bolt-on performance parts. The only solution for the amateur to the jetting problem is to pick a supplier who really understands the Weber.

Nikki carburetter:

Several Japanese cars are equipped with a twin fixed choke downdraught carburetter similar in operation to the Weber DCD type. A typical example of this type of carburetter is the Nikki manufactured by the Nippon Carburetter Company and retailed as a kit with all the parts necessary for fitting to a wide range of European cars by the British main agents, Brown & Geeson.

A typical conversion kit contains: carburetter, air filter, manifold, breather pipe adaptor (for BLMC), fuel pipes, jubilee clips and choke and throttle linkage conversions. The carburetters can be prejetted by the suppliers for the standard engine or to take into account any other performance modification carried out.

Nikki tuning points:

1 The ignition has to be advanced between 4 deg. to 7 deg. due to the position of the vacuum take off pipe on the carburetter.
2 Mixture adjustment is carried out in the normal manner.
3 The only other screw adjustment is the throttle screw.
4 The level of fuel in the float chamber can be adjusted to have a slight effect on economy by easily accessible shims – fuel level is visible through a glass screen with an etched level line.
5 For all applications there are recommended economy jets – a pack of three costs under £1.

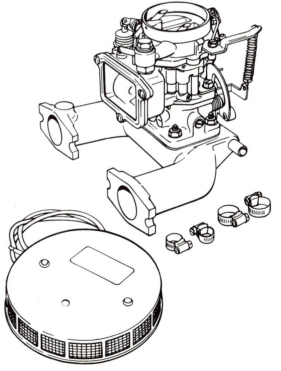

FIG 2:26 Nikki carburetter

6 Like most carburetters of this type, the Nikki is prone to freezing under damp climatic conditions – a hot air adaptor is available to overcome this problem.

Weber DCOE carburetter:

Few tuners would pretend a regard for economy in selecting the most renowned Weber performance carburetter, the DCOE twin choke sidedraught design fitted to many manufacturers' top race and rally cars. The DCOE can be infinitely finely tuned to match the requirements of a particular engine but it is not the kind of tuning process that would bear any petrol saving fruits.

Calibrating the DCOE is an expert's job demanding a great deal of experience to master the combination of jets, pumps, emulsion tubes, and chokes – all of which are variables. The home tuner will usually be left with final setting up and mounting of the unit on the car. One peculiarity of carburetters of this size is that precautions have to be taken to ensure that periodic engine vibrations

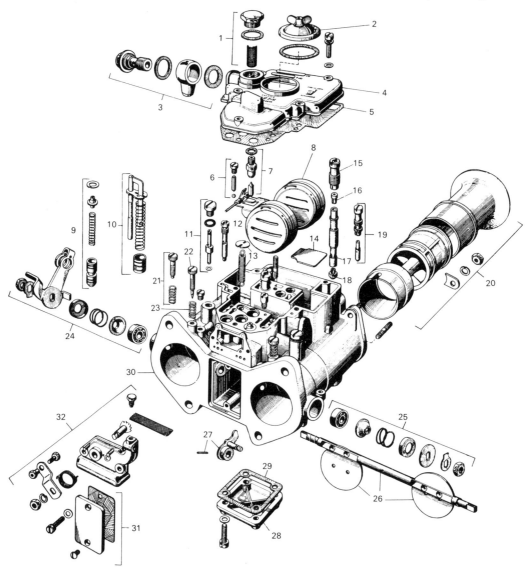

FIG 2:27   Exploded diagram of Weber DCOE carburetter

Key to Fig 2:27   1 Filter   2 Jets inspection cover   3 Fuel supply connection   4 Carburetter cover   5 Gasket   6 Ball valve assembly   7 Needle valve assembly   8 Twin floats   9 Starter valve assembly   10 Accelerator pump assembly   11 Pump jet assembly   12 Starting jet   13 Throttle return spring   14 Cover plate   15 Emulsion tube holder   16 Air corrector jet   17 Emulsion tube   18 Main jet   19 Idle jet   20 Choke and venturi assembly   21 Throttle stop screw   22 Idle mixture adjusting screw   23 Spring   24 Throttle levers and shaft fittings   25 Throttle shaft fittings   26 Throttle shaft and plates   27 Pump control lever   28 Float chamber bottom cover   29 Gasket   30 Carburetter body   31 Cover plate   32 Starter valve assembly do not upset the flow of petrol and air and that the fuel does not foam in the float chamber.

To overcome this problem the carburetters are flexibly mounted on the manifold, usually a fairly short and stubby design since the DCOEs are fairly long and there is not very much room under tuned car bonnets. The DCOE is spaced from the manifold by rubber 'O' rings which soak up the vibration. The carburetter is secured with nuts tightened down on double coil spring washers known as Thackeray washers. The nuts should be tightened down until there is a 10 thou gap between the coils of the Thackeray washer.

Tuning points for DCOEs:

1 Balancing multiple DCOEs is little more complex than the SUs. Close the throttles of both carburetters using the throttle stop screws and the linkage synchronising screw – back this one off enough to allow the adjustment of one throttle stop screw without moving the other. Turn one throttle stop screw down about one and a half turns and start the engine. Screw in the linkage synchronising screw until the throttle on the second carburetter appears to have moved the same amount. Listen with an air tube to ensure the same volume of air is entering each carburetter (some very minor differences in airflow between two chokes on the same carb can be ignored at this stage). Turn down the idle speed to normal and readjust the synchronisation if necessary.

2 The correct tickover mixture must be obtained by viewing the combustion flame in each cylinder in turn with a Colortune. On a single or multiple DCOE bank screw all the volume screws right in and then back them off two to two and a half turns. Start the engine. Apply the Colortune to each cylinder in turn adjusting each volume screw. (A good tachometer could be used to do this – it can detect the improvement in revs as each volume screw is fixed).

Reece Fish and Minnow Fish carburetters:

A carburetter that appears to work on completely different principles to all the conventional designs is the Fish, first designed in America and now modified and manufactured by two companies in Britain. It is available in kit form to suit a wide range of cars.

The Fish has a conventional float chamber with a float controlling a needle valve. Partitioned off at one end of the float chamber but bathed in petrol is a second chamber, quadrant shaped, in which a hollow arm fixed to the throttle butterfly spindle swings. The hole at the end of the arm is opposite a groove of tapering depth machined in the chamber wall. The spindle is hollow and petrol can flow through it and out of holes in the centre of the butterfly (see **FIG 2:30**).

When the swinging arm and butterfly are at the idle position the shallowness of the groove allows very little petrol through the hollow arm and spindle and into the choke tube airstream. But as the arm traverses the groove and the throttle opens, more fuel can flow through. Pump the accelerator down fast and the effect is to squeeze more fuel down the arm (a feather valve prevents back flow into the float chamber) so there is an integral accelerator pumping action which prevents flat spots.

Mixture strength is achieved by altering the attitude of the butterfly on the spindle a fraction at a time. In

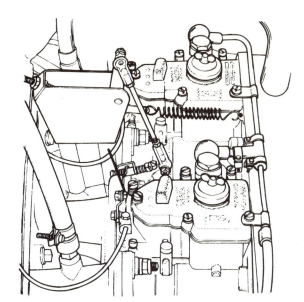

FIG 2:28 Twin Weber DCOE assembly

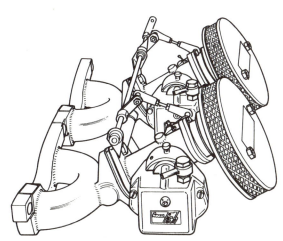

FIG 2:29 Twin Fish carburetters

other words, not changing the constant characteristics of petrol flow, but adjusting the air to suit. The idle mixture adjustment operates in the same way – a taper pointed screw controls the amount of air bypassing the butterfly in its closed position. Main jet control is by an adjustable orifice in the passageway between the swinging arm and the butterfly.

Tuning pointers for Fish carburetters:

1 Check accelerator pump type action by removing air filter, operating accelerator and watching for squirts of petrol from the butterfly orifices.
2 Fuel level can be checked by removing the inspection screw in the side of the float chamber – the petrol should be at the level of the bottom of the hole. Bend the float arm to obtain the correct level setting.
3 Set the idle to about 2000 rev/min and block off the

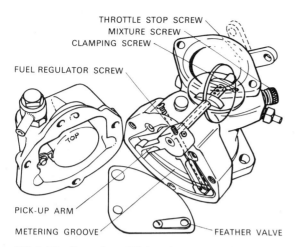

FIG 2:30 Operation of Fish carburetter

small air bleed hole in the end of the throttle spindle with a ballpoint pen tip. If the engine revs up, undo the butterfly clamp screw and **leaving the spindle in exactly the same position** turn the butterfly on the spindle to close the throttle very slightly – open it very slightly if the revs decrease. Continue this adjustment until the ballpoint pen test causes no change in the engine revs. Restore idle speed to normal.

4 Main jet setting is carried out by performance testing between 40 mile/hr and 60 mile/hr in third gear and adjusting to get the best time. The main jet is accessible under a small plug in the crook of the choke tube and float chamber – remove the plug and feel for the jet with an Allen key with the throttle fully open. Screw in to weaken the mixture – out to enrich it. Find the point at which performance increases level out and set it just slightly richer than this point, or use a Colortune.

5 Idle mixture adjustment may have been upset by this tuning – the simple screw adjuster should simply be turned back and forth to find the most even idle.

Tuning for economy with a Fish:

Claims are confused about the economy tuning benefits of this type of carburetter. In many instances correct tuning of the unit, which is after all matched completely to the needs of the particular car, has brought about good results. However they are not so startling that the outlay on the carburetter could be justified – kits cost well over £60.

Tuning for performance with a Fish:

An individually matched carburetter of this kind ought to do wonders for a conventionally factory carburated car. Again claims are confused but it is generally agreed that good tuning will result in significant power gains for small engined cars using this device – especially if a tuned inlet manifold is also used.

Amal carburetters:

Amal carburetters are among the simplest devices that can be used to mix petrol with air. Originally developed

for motorbikes and avidly used by bike racers, it was during the Mini racing boom that the use of Amals on cars came to prominence. Leyland Special Tuning do a set of four 900 series Concentric design Amals to match an eight port competitions head. Because of various peculiarities in the design of these devices they should only be thought of as competitions and performance modifications.

Amals have similarities in operation to SUs and Strombergs the major difference being that the accelerater acts directly on the air valve via a cable link. A pilot jet provides an idling mixture and thereafter the air flow and metering combination is achieved by a progressive opening of the choke tube and raising of a tapered needle in a jet orifice.

Designed as cable operated devices for bikes, the usual accelerator linkage for cars is to lead short cables to arms on a rotating bar. Individual adjustment of the arm for each carburetter is carried out for balancing purposes and main jet tuning is by changing the jets until suitable performance is achieved.

Using a Colortune tuning aid:

Tuning carburetters has become a great deal simpler since the advent of the Colortune device which is in effect a glass topped sparking plug that enables viewing of the flame colour in the combustion chamber. Tuning for various purposes can be carried out using the device.

Full instructions on operation are sent with every kit consisting of the plug, a viewing shade for use in daylight conditions, plug lead adaptors and cleaning fluid. Generally it is necessary to set the mixture to a point between rich, showing an orange flame colour, and lean, showing a bright blue flame colour. The critical point is achieved when the flame colour is a brilliant bunsen blue. However, due to carburetter and manifold design limitations it is rarely possible to achieve this accuracy across all 4, 6 or 8 cylinders. The aim should be to obtain a flame colour of bunsen blue flecked with orange on as many cylinders as possible. Should the mixture imbalance between cylinders be sufficient to cause gross tuning inaccuracy the cause should be sought and removed.

There is another useful application of the Colortune – as a fault-finder it provides clues about excessive oil usage, air leaks, valve problems and ignition system defects.

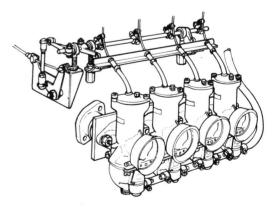

FIG 2:31 Four Amal carburetters

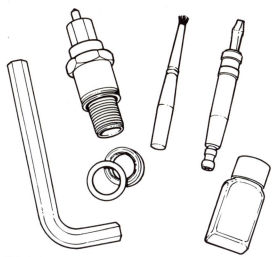

FIG 2:32 Colortune kit

Air filters and cleaners:

Tuners fitting larger engines or carburetters may be tempted to dispense with the cumbersome plastics or metal box and trunking that encloses the standard air filter or cleaner provided by the manufacturer. The apparent convenience of fitting one of the multitude of compact, chromed units sold by accessory shops as performance equipment is compounded by their attractive appearance. Avoid the temptations presented by these devices. There is sufficient road dust in the air swept into the engine compartment during normal motoring to considerably accelerate the wear of piston and rings and cylinder bores. The problems caused by dust intake will be greatly increased in competitions motoring. .

Most of the air filter accessories sold suffer from one of the following defects:

Poor filtering. The dust particles that cause bore wear are very fine – wire mesh, even if it is oil coated, cannot remove all of it.

Restriction of air flow. If a filter is incapable of passing the same volume of air as the carburetter inlet, the device acts as a restriction to the carburetter's operation. Thin pancake types can provide so little effective air entry as to create a considerable pressure drop across the filter. With the simplest type, mesh covered ram pipes, the area of the mesh itself may reduce the effective air entry by up to 10 per cent.

Poor gas flowing. Excessive convolution of the air path into the carburetter and the creation of sharp edges and turns in the inlet can cause turbulence and other air flow effects resulting in periods of air starvation or flat spots at certain air intake velocities.

There are two rules to air intake tuning: **1**, the most efficient filter medium is the standard paper element; **2**, the bigger the filter the better. However, in most cases very little improvement can be made over the standard filter for efficiency of filtration and gas flow.

In cases where underbonnet space is restricted or where it is suspected that the standard filter is not up to passing the required air flow (dynamometer or road testing is the only way of finding this out) there are three available solutions:

1 There are a very wide variety of filters and filter casings fitted to cars. A hunt round a scrapyard would provide several differing examples that would fit any particular car – aim to find the biggest.

2 A few reputable tuners sell competitions filters that retain the conventional paper element. Usually more compact than the standard units they feature improved gas flowing and a larger effective filter area. Do not accept a filter with any less effective a filter medium and design.

3 Build a filter unit to suit the car. The simplest and very effective design is to sandwich a paper filter element between two metal plates. One plate is drilled with the appropriate size of air inlet hole and, if necessary bolt holes for flange mounting. The second plate is simply drilled with the hole for the existing centre air filter bolt or to take long bolts from the carburetter flange. Whatever arrangement is practical, ensure that there are no leaks between the filter and the carburetter intake and avoid sharp turns in the air path. The edges of flanges and plates should be blunted or even better, radiused.

2:3 Cylinder head modification

The cylinder head can be modified by the home tuner with the most far reaching effects on the car's performance and economy. Standard cylinder heads will suffer from any or all of the following defects that detract from the efficiency of the petrol/air combustion process:

Uneven combustion chamber volumes. Standard production cylinder heads are manufactured at speed using tools that eventually wear. Quite apart from the inaccuracy in combustion chamber volume that may result from the head casting process any cutting or other work carried out on the size or finish of the head or the chambers will introduce small differences in the volume and area of the chambers.

Poor matching of head to block and manifolds to head ports. The same casting inaccuracy and tool wear that affects the head also affects the critical dimensions of the block. The two sets of errors can combine to produce considerable mismatching of the chambers to the cylinders. At best the chambers will not be large enough to cover the full cylinder orifice – at worst there can be overlaps between the edge of the cylinder and the chamber. The same fault is evident at the interface between the head and manifolds – the ports are rarely the

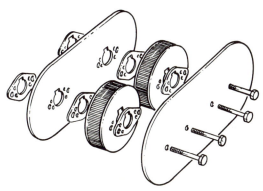

FIG 2:33 Home-made air filter

same size or, in minute detail, shape and casting and machining inaccuracies very often mean that they are in the wrong relative locations for accurate mating. The steps this causes between the parts seriously affects the gas flow.

Rough finishing. Few production cylinder heads are produced with a finish that is conducive to the best combustion efficiency. Rough port and chamber surfaces create drag and turbulence in the petrol/air mixture that prevent the maximum possible charge entering the combustion chamber and may also affect the evacuation of exhaust gases. Rough projections can heat up and cause uneven combustion, pre-ignition and running on problems.

Head warping. Although head warping may occur as a result of abuse of the engine a slight degree of warp is often found on new heads owing to the fact that stresses set up in the metal during casting are not relieved prior to machining of the head surface. The stresses are relieved by age and heating – after a few miles the head that was true when it left the machine shop assumes a warp (this is a problem that can also occur with blocks).

All these faults can be eradicated with the procedures described in the following sections. Head modification can be carried out with a view to producing an economy car retaining the standard compression ratio or a performance car with an engine extended to the maximum possible safe and efficient combustion limits. The practical aspects of the work involved are the same.

Preparing the cylinder head:

Remove the cylinder head from the car and take off removable parts like heater pipes and taps, the thermostat and housing and the studs used to secure manifolds. It is still useful to keep the valve mechanism in place at this stage.

Decoke the head and valve faces removing all traces of carbon, old gaskets and gasket cement. This should be carried out using a small wire brush in an electric drill – scrapers may be used to remove stubborn deposits. Do not bear too heavily on the wire brush as this can scratch the head metal unnecessarily.

Measuring the cylinder head:

Cylinder head modification has to be carried out to within very fine volume tolerances and it is also necessary to ensure that the metal of the cylinder head is thick enough to withstand the removal of a considerable amount of metal in a few key areas.

Determining combustion chamber volume:

An analyst's burette, available from chemical equipment suppliers, is necessary to measure the volume of the combustion chamber. An additional useful piece of equipment must be made. It is a clear Perspex blanking plate large enough to cover the combustion chamber with a $\frac{1}{4}$ inch hole drilled at its centre. Use the following procedure as a preliminary determination of the largest combustion chamber:

1 Insert the spark plugs to the correct tightness.
2 Support the cylinder head so that it is dead level on the work bench – use a spirit level for this critical operation.

3 Support the burette vertically above the first combustion chamber to be measured.
4 If there is any doubt about the sealing of the valves in the combustion chambers wipe the sealing rims with a smear of heavy grease such as Shell Retinax.
5 Seal the blanking plate over the combustion chamber with heavy grease.
6 Fill the burette to just below the zero level with a mixture of paraffin (5 parts) to clean engine oil (1 part). Clear air locks in the burette tap by running out a small amount of the mixture onto a rag. Take the burette reading.

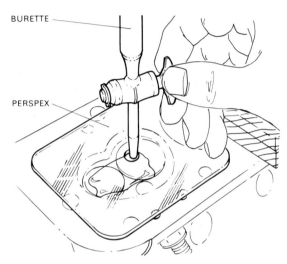

FIG 2:34 Measuring combustion chamber volume

7 Slowly run mixture from the burette into the combustion chamber through the hole in the blanking plate ensuring that, as a liquid level rises in the chamber, all bubbles float out of crevices like the spark plug electrode gap. Fill the chamber to the level of the blanking plate hole.
8 Note the amount of mixture that the burette has delivered. This is the combustion chamber volume.
9 Carry out this procedure on each of the chambers.

Make careful notes of the volume of each chamber – all must be modified to a greater or lesser extent so that they eventually have an equal volume to that of the largest chamber found in this test.

Determining the metal thickness:

On every head there are crucial areas where the removal of metal may reduce the wall thickness of oil galleries or water channels to the point where stress or corrosion may break the cylinder head face (thus rupturing cylinder seal, causing oil and water mixing or breaking the gasket) or the combustion chamber wall. It is usually quite difficult to gauge the thickness of the metal around combustion chambers and at the head face – every cylinder head, even on the same model, varies. However, it is sometimes possible to use vernier calipers or to make a reasonably accurate assessment of thickness from probes (a bent wire is useful) inserted into the oil and water drillings on the head face. There is no general guide to these measurements; each one is up to the tuner's ingenuity.

The only rule is that when a head has been modified and the head face machining has been carried out there should be at least 75 thou of metal at all points throughout the head. Experienced tuners may be able to skim metal to within finer limits in certain circumstances but the home tuner must always work to the limit above.

Determining volume due to cylinder head gasket:

Make a simple clamp of two pieces of iron or steel at least $\frac{3}{8}$ inch thick drilled with two holes so they can be secured together by high tensile bolts. Clamp a piece of the old gasket material between the two pieces of metal and tighten down the bolts to the torque indicated for cylinder head bolts in the workshop manual. Measure the total thickness of the sandwich with a micrometer. Undo the clamp, remove the gasket material and retighten. Measure the clamp thickness. Subtract the second figure from the first to obtain the gasket thickness under compression.

The combustion chamber volume due to the cylinder head gasket can be calculated by multiplying the gasket thickness by the area of the combustion chamber at the head to gasket interface. This can be estimated by tracing the shape of the chamber onto square ruled graph paper and counting the number of squares inside the outline – work out the size of part squares intersected by the outline to add to the number of whole squares. In practice it is accurate enough to count the number of intersected squares and divide by two to find the peripheral area.

Other volumes:

To determine combustion chamber volume on bowl-in-piston designs and where there are volume variations

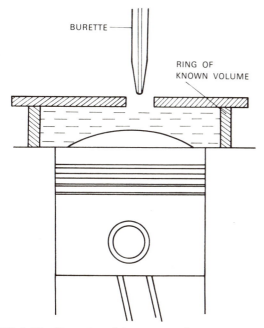

BURETTE

RING OF KNOWN VOLUME

FIG 2:35   Measuring piston crown volume

like valve recesses in the piston crown or raised crowns, measurement of volumes in the block are carried out in a similar way to those in the cylinder head. With the engine in the car or out, the top face of the block has to be settled completely level.

Smear a little heavy grease on the cylinder bore and bring the piston to be measured up to top dead centre. For raised crown designs (and those engines on which the piston face comes very nearly flush with the block face) use a ring of metal of known depth sealed to the block face around the bore with grease. Cover the cylinder or the ring with the Perspex plate. Carry out volume measurements in the manner described above. The raised or depressed crown volume can be found by subtracting the known volume of the ring from the volume found by burette. In the case of a raised crown piston the volume will be a minus quantity to be taken from the combustion chamber volumes found in the head measurements taken previously (see **FIG 2:35**).

Calculations:

A number of vital tuning calculations can be made from the measurements made in the previous sections. In the following equations the values obtained are abbreviated as follows:

V = the swept volume of a cylinder (this can be obtained from the workshop manual for an unmodified engine otherwise it has to be calculated; remember that reboring will alter this figure)

r = bore radius (diameter divided by 2)

s = stroke of piston

c.r. = compression ratio

C = combustion chamber volume

G = volume due to gasket

P = volume due to raised piston crowns etc.

A = area of combustion chamber at head face

1 Swept volume:

$$V = \Pi r^2 \qquad (\Pi = \frac{22}{7})$$

2 Compression ratio:

$$c.r. = \frac{V + C + G + P}{C + G + P}$$

3 Combustion chamber volume to achieve a specific compression ratio:

$$C = \frac{V}{c.r. - 1} - (G + P)$$

4 Head thickness to be removed (X) to achieve given chamber volume (Y):

$$X = \frac{C - Y}{A}$$

Working on the cylinder head:

After the measurements described in the preceeding paragraphs the valves can be removed from the head using a valve compressor tool. The valve mechanism parts should be stored in such a way that the springs, collets, and so on for each valve are kept together and that the valve for each position is clearly identified. Swopping parts can have disastrous results.

FIG 2:36   Improvised head holding bolt

Do not remove the valve guides – these have a useful function in locating an old valve used to protect the valve seats during combustion chamber work. An old valve can be modified for this task by rounding down its end face to leave just the narrow seat face intact.

Mount the cylinder head firmly on the workbench which should be as clean as possible. A good head mounting can be made by Aralditing long bolts into the threaded shanks of spark plugs (with insulators and electrodes removed). The bolts can then be threaded into the head and their free end clamped firmly into a vice.

Fix an old cylinder head gasket in place on the head face ensuring that it is properly located by passing bolts or pegs through at least three of the bolt holes. Scribe round the combustion chamber orifices in the gasket. This is the area within which all modification work must be carried out. Extending the chamber beyond these limits will cause gasket problems. Greatly reducing the flat 'squish' area over the piston causes trouble, too.

Make a template out of Perspex to the shape required for the finished combustion chamber. When deciding on

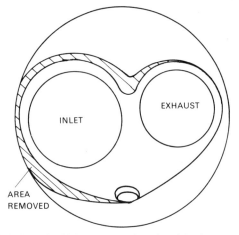

FIG 2:37   Modifying combustion chamber shape

a combustion chamber shape never strike out without some advice. Most of the tuning specialist periodicals run regular articles on cylinder head modification which include recommended patterns – watch out for articles on the particular car being tackled. Improving breathing around the inlet valve area is a common modification (see FIG 2:37). There are two alternatives to the use of these patterns. First it may be possible to persuade a tuning shop to show a factory modified head so that an outline can be traced or drawn. Secondly very good results can be obtained by simply ensuring that every chamber is modified to the shape of the largest one in the head. This is because there is usually very little wrong with the standard combustion chamber shape – it is the finishing and balancing that is sub-standard.

Removal of metal to obtain the required shape is a painstaking and slow process using miniature grindstones on a flexible extension to a motor or electric drill. Spherical and cylindrical stones are the best answer to working inside combustion chambers as there is little possibility of making disastrous slips that could score the metal too deep for removal of the mark. Great care should be taken in using pointed and conical stones to reach more acutely angled corners of the chamber. Remember to protect valve seats with an old valve while working on the head. **Wear goggles or a visor to protect your eyes.**

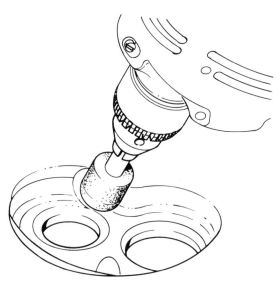

FIG 2:38   Using a grinder to modify a combustion chamber

Once the required profile has been achieved the remainder of the work can be carried out using various grades of emery paper or cloth stuck onto small rubber sanding discs. The final finish need not necessarily be an absolute mirror polish. The aim should be to remove all visible surface imperfections like casting marks and grinder scores – these are the points at which carbon will build up, giving rise to pre-ignition and running on problems.

Before giving the chambers a final fine grade polishing, lap the valves into place roughly with grinding paste using the valve suction grinder. Use heavy grease to ensure a proper seal and check the combustion chamber volume very thoroughly taking at least two readings on each chamber and averaging out the result to ensure accuracy. Compare the volumes obtained to see that they are accurately balanced. Carrying out the final polishing will not significantly alter the figures obtained.

Valves, valve guides and valve seats:

Up to now the head has been modified with little consideration to the valves. But these are just as crucial to gas flowing as the rest of the cylinder head. Worn valve guides must be replaced, in which case the seats have to be recut to centre the valves in the new positions.

Good valve guides can be removed by the home tuner for modification provided that a few simple precautions are observed:

1 The guides must be carefully and gently marked to ensure they go back in the right holes and in exactly the same position. This should be done by lightly scoring block and guide with the reference mark and marking the guide with a number identification code. Do all marking at the valve spring end.

2 Drive out guides from the top using a special drift made from a bolt of the same diameter as the valve shank with its head reduced to the outside diameter of the guide. Insert the drift in the guide and lightly drive out the bolt head with the guide on it. It will be possible to see the extent to which the valve guide protrudes from the guide boss into the inlet or exhaust port by the carbon marking. Taper this protruding end back for a distance up to $\frac{1}{4}$ inch – bear in mind that the guide may protrude a lot more when the ports are modified (see **FIG 2:39**).

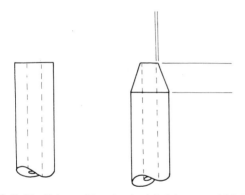

FIG 2:39 Valve guide: standard (left), tapered (right)

In modifying the valve itself great care should be taken to ensure that the stem, the seat and the retaining clip groove are not damaged in any way. Use a grinder, bench mounted type, to narrow down the seat area by reducing the top face edge to a fine radius. Similarly lower the profile of the underside of the valve face increasing the radius of the head to the stem but guarding against stem damage (see **FIG 2:40**).

Polish the valves by mounting them in the chuck of a bench mounted drill and using emery paper while they rotate.

Valve seats must be carefully radiused into the roof of the combustion chamber but here the great difficulty is in avoiding valve sealing problems. Any slight imbalance in the radius from one side of the valve seat to the other can mean the cutting of a new seat to get the valve to seal. One way is to use a rubber washer of slightly larger diameter than the valve seat with fine emery cloth stuck to one side of it. Make sure the hole in the washer is a reasonable interference fit on the stem of an old valve. The valve fitted with the washer can then be passed through the valve guide. An electric drill on the spring end will rotate the valve and emery covered washer to radius the seat area. The radius can be extended at least as far as the seat by pulling on the rotating assembly. Do not exert too much pressure or this will destroy the seat.

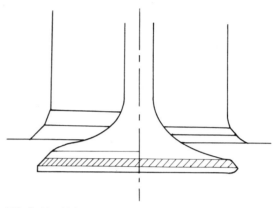

FIG 2:40 Valve and seat: standard (left), modified (right)

Ports:

Mark the manifold face of the head with the gaskets to be used in the same way as the combustion chambers. The scribed line is the maximum permissible limit for metal removal. The rule for ports is almost the same as for combustion chamber modification – the profiles of each port must be the same. Do not waste time on carrying out exhaust port modifications; the power or economy gains to be made by doing this are extremely small.

Unfortunately volumetric measurement of ports is not practical – the size must be estimated by using suitably shaped metal templates inserted from the manifold face and from the valve throat. It is not normally possible to make these two templates meet to check the area of the bend. This has to be felt carefully using light finger strokes to check that the radius of the turn and smoothness of finish is comparable to other ports. Smooth over and around the valve guide boss bearing in mind the shape already imparted to the guides themselves.

Polishing of the ports is carried out by inserting rotating pieces of emery paper snagged in a fork cut in a length of metal rod fitted on a drill chuck (see **FIG 2:42**).

Matching manifolds to the head:

The main problem occurring in matching inlet and exhaust manifolds to the cylinder head is that very few cars have any really positive location of the manifolds in their standard form. To ensure that manifolds cannot

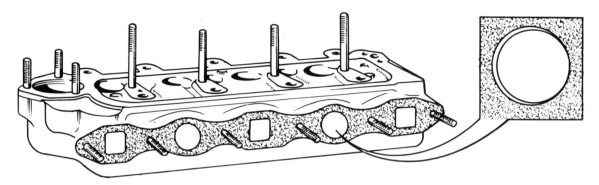

FIG 2:41   Marking gasket size on port face

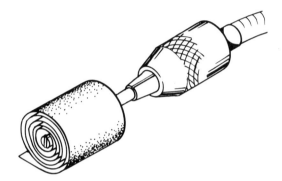

FIG 2:42   Port polishing tool

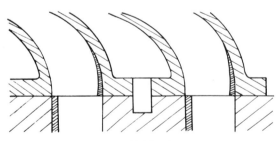

FIG 2:43   Matching manifold and ports

drift away from a carefully prepared and matched position it is a good idea to fit dowel pegs between the block and manifold flange. A $\frac{1}{4}$ inch deep drilling is enough to locate the pegs which can be made from steel rod of say, $\frac{3}{16}$ inch diameter, but this should be carried out at a suitably meaty part of the head and manifold flange.

Once the manifold is properly located on the head it is possible to determine the amount of matching necessary. Matching is simply the process of removing the steps in the flow that occur as the result of a manifold orifice

poorly positioned on a head orifice (see **FIG 2:43**). The process to find the offset is tedious but quite simple.

Use a suitably sized piece of stiff paper stuck lightly onto the head face over the locating pegs – holes can be cut to accommodate these. Rub the paper with a soft pencil to show up the port outlines. Remove the paper from the head and use a scalpel or razor to cut out the holes marked. Offer up this paper template to the mani-folding, remember that the reverse side must be used. Locate the template as accurately as possible. The areas revealed as overlaps on either the head or the manifold can be removed using the same stones as for head modification. It may be necessary to trim gaskets to fit the matched ports as well.

Machining the head:

With the head in this state of preparation it is now possible to raise or restore the compression ratio. For economy tuned engines it is necessary to return the compression ratio to the original value before all the metal was removed in balancing and polishing. Raising the compression ratio could mean a need for a higher, uneconomic, grade of petrol.

There are no fixed guide lines for the home tuner in determining the possible compression ratio. However, unless several examples of tuners doing otherwise with a particular car can be found, it is not wise to stray beyond a compression ratio of 10:1.

There are two ways to measure the amount of metal to be skimmed off the head to achieve a desired compression ratio. The first is to run the amount of paraffin/oil mixture indicated by working equation **3** into the com-bustion chamber from a burette. Then, using a dial gauge, the level of the liquid below the head surface can be measured. This distance plus another 12 thou to account for the tendency of the liquid to rise up to the dial gauge needle is the amount to be skimmed off (see **FIG 2:44**).

The second method is only accurate if the combustion chamber is reasonably straight sided. Measure the area of the combustion chamber using the graph paper method. Subtract the desired volume from the known volume and divide the area found into the result (equation **4**). The answer is the amount to be skimmed.

Specialist machine shops will skim off the head (also curing slight warping at the same time) and if necessary recut the valve seats. Countersink all bolt holes in the head and hand file or grind the sharp edges of the combustion chambers.

Grinding in the valves in the last major job before reassembling the completed head. But before bolting it all together make absolutely sure the head is clean by thorough washing in paraffin and then lots of detergent and water. Dry thoroughly and coat with a thin layer of oil.

Raising compression ratios on flat headed engines:

Engines with a bowl-in-piston design of combustion chamber have a flat faced cylinder head – the modifications carried out are restricted to radiusing and polishing of the valve seat area and port and manifold work. Any work on the pistons, which are in any case more precision components than standard cylinder heads, would be useless as the metal removed would lower the compression ratio. The big change that can be made to affect the performance is to raise the compression ratio by fitting specialist high performance type pistons or the smaller chambered pistons from lower capacity engines in the range.

The use of high performance pistons:

Apart from their use in high compression ratio flat headed cars there are a wide variety of high performance pistons available from specialist manufacturers like Hepolite. There are a number of reasons why uprated pistons are an advantage on a performance car. They are able to cope with higher engine revs, they will operate without stress at higher pressures and temperatures and they may have different types of oil or pressure rings. Another useful characteristic is that they are available in a wide range of sizes for various cars so that the larger piston requirements of rebored cars can be accommodated.

2:4 Camshafts

Changing the valve timing and lift characteristics by using a sports or other high performance cam is not an economy measure. The purchase of the cam (about £10 to £40) is sufficient to dissuade economy tuners. Ideally camshaft changes should be coupled with uprating of the valve gear as a whole. Stiffer valve spring sets (in some cases double springs) are recommended by most camshaft specialists and it would be advantageous to renew the rocker gear on an older car, perhaps investing in lightened high performance components.

What do sports camshafts offer? Such a camshaft alters the valve timing, generally giving greater overlap between the opening of the valves on different cylinders and longer opening periods. These changes usually mean a sacrifice of engine tractability at lower revolutions which is offset by a gain in power at higher revolutions. Once again the tuner has to have a clear idea of what is expected of the car. The only general rule is that the camshaft for fitting to a road car should have a specification that is little more than that the manufacturer provides for the higher powered or sports models in a range.

Harmonic cams are the simplest type and are widely fitted to production cars. They are easily made by mass production methods and have a simple geometric profile.

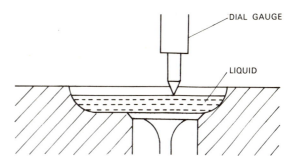

FIG 2:44   Measuring the amount to be skimmed off the head

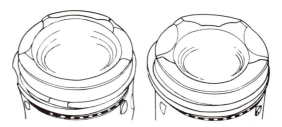

FIG 2:45   Ford pistons: 1600 (left), 1300 (right)

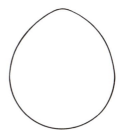

FIG 2:46   Harmonic cam profile

FIG 2:47   Polydyne cam profile

Some production high performance cars have a different type of cam that is also available as a performance accessory for some cars. This polydyne type has a far more complex profile but the main feature of the design is that it takes up all the tolerance slack in the valve gear so that the valve lifts (usually higher than a standard cam) immediately it is required to.

Before making changes to the cam precautions should be taken to ensure the accuracy of the car's standard timing mechanism. Full benefits of a performance cam can only be achieved by spot-on valve timing. Manufacturers may be able to provide special keys or wedges or replacement sprockets with vernier mountings to make minute timing adjustments possible.

2:5 Lightening, toughening and balancing

It has already been mentioned that lightened rocker gear is available for performance tuning. Other lightened parts widely available for competitions use include lightened flywheels, crankshafts and connecting rods. These are all components of a balanced system designed for very high revolutions.

Lightening and balancing is not a home tuner's job. Specialist tuning shops will carry out the work. Additional performance advantages may be obtained by toughening some parts. The crankshaft in particular accepts a great deal of the strain in performance tuning and there are a number of similar commercial processes available to strengthen this vital component. Tuftriding and nitriding are two common names for the process.

2:6 The cooling system

In the section on fitting a larger engine the problems associated with cooling the larger unit were discussed. Similar problems arise in tuning a standard engine for economy and in creating a performance car. However, in these two instances the approaches are completely different. In an economy car the main aim is to raise or stabilise the engine temperature. In the case of a performance car it is necessary to provide for better heat dissipation as the tuned engine will generate higher waste heat output under some conditions.

Economy tuning of the cooling system:

Most cars are over-cooled in standard form. The engine takes longer than should be required to reach a stable temperature, thus the choke is used too much to maintain even running and inefficiencies in petrol usage arise from the poor vapourisation of improperly atomised petrol in the inlet manifold.

The simplest way to raise the engine temperature quicker is to reduce the efficiency of the standard cooling system. In most cases this can be accomplished by cutting down the amount of air flow through the radiator. Special blinds or mufflers can be purchased which, fit in front of or behind the grille – sophisticated types can be adjusted from inside the car. These are not necessary; it is sufficient to blank off the grille with cooking foil, metal sheet, plywood or similar reasonably weather resistant material. Start by blanking off the entire grille – if this results in overheating problems, reduce the area covered until the best results are obtained.

When a blanking plate is fitted, make absolutely sure that it does not cover the inlet for fresh air ventilation or heater systems.

The second most effective means of improving the rate of warm up is to fit a winter thermostat, that is a thermostat which opens to allow full water circulation through the block at a higher temperature than standard. Some manufacturers recommend a high temperature thermostat for their cars – for example the standard Mini has an 82 deg. C thermostat (this is stamped on the end

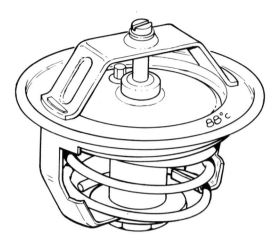

FIG 2:48 Thermostat showing temperature grade

of the device) and it is possible to fit one that opens at 88 deg. C. Even if there is no high temperature thermostat recommended it should be possible to obtain one that will give a 5 deg. to 8 deg. C rise in opening temperature. For many cars it is possible to use this winter thermostat all the year round.

The use of electric or feathering fans is discussed later. These devices are expensive and offer such small improvement in petrol economy that it is not worthwhile fitting one for mile/gall gains alone.

Performance tuning of the cooling system:

The higher heat outputs of a performance tuned engine may demand the fitting of a more efficient radiator. A radiator from a higher powered model in the same range may be used – new or scrap parts are suitable. It may also be possible for a radiator specialist to rebuild a standard

FIG 2:49 Electric cooling fan

radiator with a core that has greater dissipation characteristics.

In fact in the majority of cars tuned for road use the spare heat dissipation capacity of the standard system is sufficient to cope with the heat output.

Generally, the problem is still one of overcooling. But in the case of the performance tuned car it is not a good thing to make semi-permanent changes to the heat characteristics of the heating system like changing the thermostat or blanking the grille. These may prove too restricting when the performance is used; a more flexible system is called for.

If the car does not already have one, flexibility is provided by fitting an accessory electric fan. There are two electric fan kits on the British market – the Kenlowe, a small fan, long bodied motor design, and the slimmer, wider bladed Wood Jeffreys type.

The advantages of both these units are that they reduce the power loss due to the useless churning of the standard fan at high speed when the ram air is sufficient to cool the engine. At all speeds they also reduce engine

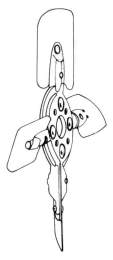

FIG 2:50   Aerofan variable pitch fan

noise levels. They are supplied as kits with all electrical connectors and wiring – both have a range of adjustment to set the temperature at which the fan cuts in.

A cheaper solution is to use the one variable pitch fan available, the Aerofan. This has sprung blades which reduce in pitch as engine speeds increase.

As mentioned above in the economy tuning section, the use of an electric or variable pitch fan can reduce fuel consumption. Some fuel is saved by electric fans in reducing the warm up period while all the devices reduce power wastage. The gain is not sufficient to justify use of these kits as an economy aid alone. Tuners who consider that the gain in power, and particularly the reduction in engine noise, are valuable may find these sufficient reasons to fit fan units.

2:7 Tuning the exhaust system

Just like the restrictions on performance imposed by the design of mass production carburetters, standard exhaust

FIG 2:51   Typical tuned exhaust system

manifolds are often poorly constructed from the point of view of efficient disposal of the engine's waste gases. Ideally it is possible to create an exhaust manifold system that actually sucks the exhaust gases out of the cylinder when the exhaust valve is opened. Many special tuning exhaust systems come close to this ideal by sophisticated juggling with the lengths of the individual branches from each cylinder and the arrangement of the point at which the branches join into a single outlet tube.

The only modification that the home tuner can carry out is to fit one of these systems obtained from a specialist supplier or from another higher powered model in the same range. Where this is possible using cheap scrap parts it is a valuable economy modification.

The silencing system has to conform to legal requirements which do, in this instance, conflict with performance aims. The noise of an unsilenced but efficiently tuned system would be unbearable. Specialists should be consulted about the most efficient silencing system for a particular car.

Do not be taken in by advertisements claiming economy and performance gains from devices that bolt on to the end of the tail pipe. These can have no affect whatsoever on the crucial gas flow characteristics of the manifold itself and claims to accelerate exhaust ejection are nonsense as the exhaust gas velocities of a car at speed far exceed any assistance possible from the airstream of the car.

2:8 The lubrication system on performance cars

The greater the stress placed upon components and the higher the heat load dispersed through the cooling system the greater the reliance upon an efficient lubrication system. High performance cars must have lubrication systems in peak condition. This means the tuner must ensure that key components of the system like the oil pump are in good working order before putting a performance modified car on the road.

Particular points to watch are that the rebuilt engine is sufficiently well lubricated during assembly so that it can run for a few seconds or minutes before an adequate supply of sump oil reaches the bearing surfaces; that parts like the rocker gear which have internal oil channels must be assembled with the appropriate oil galleys correctly matched through the head, rocker post and shaft; that the filter must work effectively and it is wise to supplement its action with a magnetic drain plug.

At one time there was a craze for fitting oil coolers. These expensive items were often doing more harm than good, cooling oil to below its most effective operating

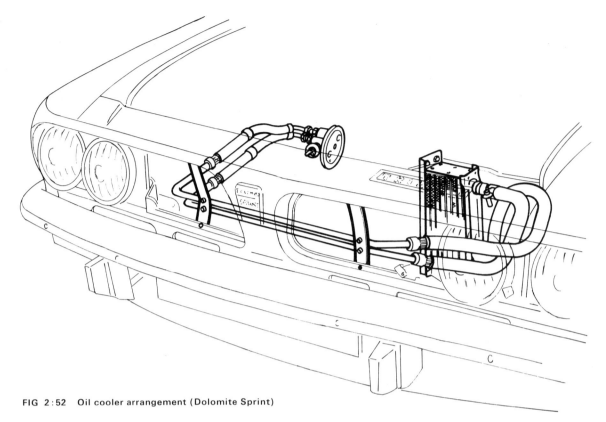

FIG 2:52   Oil cooler arrangement (Dolomite Sprint)

temperature. In fact oil functions best at about 100 deg. C and modern multigrade oils can operate at well above this level. Few cars will achieve this temperature consistently during normal road motoring. The few remaining occasions when an oil cooler becomes an advantage are in competitions use when constant high engine revs and relatively low ground speeds combine to produce an unusual heat build up – both oil and water cooling systems react violently to this kind of treatment. The solution for a car used in this kind of event as well as on the road, is an oilstat plumbed into the oil cooler system. This device acts just the same as the thermostat and will only allow passage of oil through the filter above a certain preset temperature.

The only concession that the tuner should make to hard use in road or competitions motoring is to make oil changes more frequently. A suggested interval for oil changes on a modified car of conventional design is 4000 miles – on a transverse BLMC engine changes could be made every 2000 miles.

Air-cooled engines are a special case, as the oil plays an important part in the cooling. Anyone carrying out more than mild tuning on a VW Beetle, for example, would be well advised to fit an oil temperature gauge and to consult a tuning firm with experience of the particular model in question about possible modifications to the lubrication system.

Very high performance modified rally and racing cars are often fitted with a system of lubrication known as dry-sump. This means just what it says – oil is drained out of the sump (often modified to improve oil collection) to a large capacity oil tank from where a powerful pump distributes it around the system again. There are several advantages to this system. Firstly, more oil can be carried to accommodate the extra oil consumption of a high performance car and the greater stresses on the lubricant. There is no surge and churning of oil in the sump and therefore there is little of the aeration and frothing that cuts down on the lubrication efficiency. Lastly a greater volume of oil has a greater capacity to conduct heat away from the engine. It is not a system that could be recommended as a modification for the road-going tuned car.

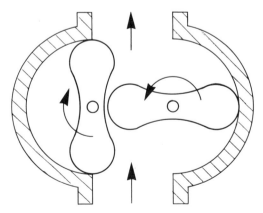

FIG 2:53   Roots-type supercharger

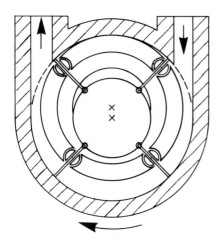

FIG 2:54 Shorrocks vane-type supercharger

2:9 The ultimate in tuned engines?

Many specialist tuners and some car manufacturers are coming round to the view that the most effective use can be made of an engine if a system of tuning called turbo-charging or supercharging is used. They speculate that an engine designed to operate with one of these types of systems will be able to overcome the stringent pollution regulations currently hampering car design and provide the same power output as much larger and more expensive engines with potentially better fuel economy.

Superchargers and turbochargers are not new. Super-chargers were quite widely used on sporting cars like Bentleys and Bugattis during motoring's formative years before the Second World War. Turbocharging has only more recently been applied to cars, but is well established and widely used on truck engines.

Both systems employ an air pump or compressor to raise the pressure in the engine's inlet tract so that more air/fuel mixture is forced into the cylinders. Roughly speaking, the more mixture the engine can breathe the greater its power output.

The difference in the two systems lies in the way the compressor is driven. In supercharging, the compressor – the Roots type, with a pair of lobed rotors (see **FIG 2:53**) and the eccentric vane Shorrocks type (see **FIG 2:54**) are the commonest – is driven mechanically by belt, chain or shaft from the engine crankshaft. In a turbocharger, a centrifugal compressor is turned by a turbine driven by the engines exhaust gases (see **FIG 2:55**), harnessing energy which is otherwise lost.

The driver of the tuned car will find it absolutely necessary to know more about how the lubrication system is functioning, however. The best way to obtain this information is by fitting an oil pressure gauge so that any flagging in pressure is immediately detectable before damage occurs. Lest there be any lingering doubts about oil cooling during hard driving an oil temperature gauge can also be installed.

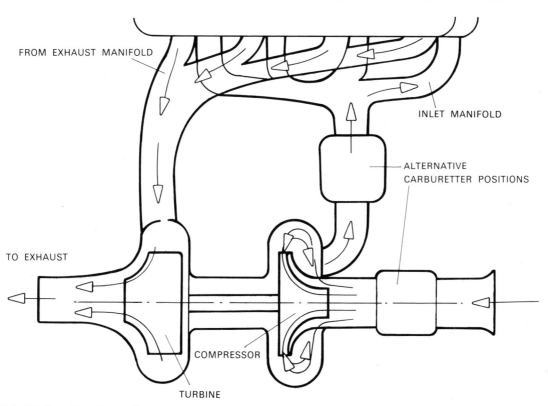

FROM EXHAUST MANIFOLD

INLET MANIFOLD

ALTERNATIVE CARBURETTER POSITIONS

TO EXHAUST

COMPRESSOR

TURBINE

FIG 2:55 Principle of turbocharger operation

The turbine and compressor unit is very compact and well balanced, and runs at speeds of up to 90,000 rev/min.

These techniques provide the nearest thing so far devised to instant bolt-on power because large increases in power output can be obtained with few additional modifications. In fact, paradoxically, the engine itself has to be slightly detuned to cope with the extra gas through-put. The pressure applied to the intake system – around 10 lb/sq inch in a typical road-going conversion – effectively raises the compression ratio so that the actual ratio has to be lowered in compensation.

While the principle is simple, the development work involved in producing a reliable and untemperamental conversion is not always so straightforward. But the comparatively small number of specialists who have persevered in this field have been able to demonstrate some impressive results. Turbocharging in particular is beginning to appear as a production option, its potential limited only by considerations of cost.

CHAPTER 3

Tuning the ignition system

3:1 Conventional ignition systems
3:2 Practical ignition tuning
3:3 Static ignition timing
3:4 Stroboscopic ignition timing

3:5 Alternative ignition systems
3:6 Tuning summary – for economy
3:7 Tuning summary – for performance

3:1 Conventional ignition systems

In tuning the car so far the emphasis has been on introducing the correct mixture to the combustion chamber, at the right moment, ensuring the optimum conditions for smooth combustion and getting rid of the waste in the most efficient manner. Of equal – and some would argue, greater – importance is the progress of the combustion itself.

The means by which ignition is achieved is the discharge of high voltage (high tension) current across a small gap between two electrodes – the result is a high temperature spark. The electrodes are introduced into the combustion chamber as the spark plug. The spark plug electrodes are usually designed to protrude into the chamber provided clearance of valves and pistons allows.

The flame of ignited petrol-air mixture takes time to spread in an ever widening front throughout the mixture. Because of this time lag in the development of maximum pressure from the burning of the mixture the spark must be timed to occur an instant before the piston reaches top dead centre (tdc) on the compression stroke. In this way maximum pressure is reached shortly after the piston is past top dead centre.

The rate of spread of the flame front in the mixture (and therefore the amount of time the spark should be made before the piston reaches tdc) varies according to four factors: **1** the ratio of petrol to air in the mixture; **2** the density of the mixture (therefore its temperature and pressure); **3** the design of the spark plug and the combustion chamber; **4** the strength of the spark. For any given engine these factors are relatively constant. But there is one constantly varying factor and that is the speed at which the piston approaches tdc. Therefore the method by which the spark is generated has to include some means of advancing the spark timing to ensure that

maximum combustion energy is achieved shortly after the piston passes tdc.

Another varying factor is the load on the engine. Light loads at high speed demand an advanced spark; heavy loads and lower engine speeds require the spark to be retarded. So there also has to be a means of compensating for these conditions in the spark timing mechanism. The factor which varies, albeit crudely, in sympathy with the engine conditions is the depression in the induction manifold. This depression or vacuum is harnessed to a spark advance mechanism which is a part of the conventional distributor – in fact it is the distributor which is the key to all these timing functions so it is worth examining its action in some detail.

The distributor's action:

Current is provided to the ignition coil primary windings through the ignition switch and the earth return is via the contact breaker points in the distributor. Parting the contact breaker points switches off the current to the primary winding causing a collapse of the magnetic field around the coil core. The collapsing lines of force which form the field intersect the coil's secondary winding and a high tension lead carries the resultant high voltage impulse to the centre terminal on the distributor cap which is connected inside the cap to a brush bearing on the rotor arm. The high tension impulse flows along the rotor arm conductor and jumps a narrow air gap to the nearest of the terminals placed radially around the cap (one for each spark plug). Thus the distributor has a dual switching role, acting on the primary winding current to produce the magnetic field collapse and subsequently ensuring that the spark impulse is directed to the right plug.

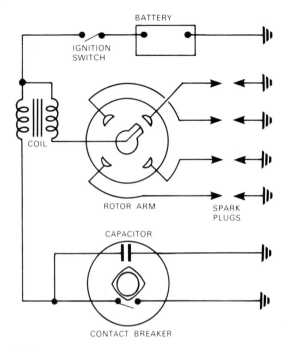

FIG 3:1   Ignition system circuit diagram

The rotary action necessary for this switching is provided by the distributor drive shaft which is driven from the engine's camshaft or crankshaft. On almost all engines the distributor shaft rotates at half crankshaft speed.

The rotor arm is attached (usually a push fit) on the end of the distributor cam which has as many lobes as the engine has cylinders. The cam, which performs the contact breaking action, has a few degrees of freedom to rotate concentrically about the drive shaft to which it is attached.

The contact breaker:

The contact breaker is mounted on the distributor base plate which may be sited above or below the automatic or centrifugal advance mechanism. Bosch, Motorcraft/Autolite and Lucas units have the base plate above the automatic advance – some AC Delco (Vauxhall) and Marelli (Italian cars) distributors have the base plate mounted below the automatic advance weights.

The points themselves are small pads of an especially hard tungsten alloy. One is on a steel mount adjustably positioned by a screw fixing to the base plate. The other contact is on a short length of spring steel and moves with a hard plastics cam follower or heel which bears on the cam. As the cam rotates each lobe pushes the two contact points apart and breaks the current to the primary winding.

The gap created when the contacts open is a vital factor for engine performance. It contributes to the efficiency of the contact breaking operation itself but more important is its effect on the length of time given for the coil's magnetic field to recover its strength (the dwell angle, discussed later).

For normal motoring, with economy in mind, contact breakers are adequate devices. They are cheap to replace, reliable and easy to maintain and they work reasonably well despite their relatively simple design. Their service life is not long – most experts advise replacement at no longer than 5000 miles intervals but the gap should not need adjustment more than once or twice in this time. Contact breakers are generally designed so that wear on the cam follower compensates for wear of the points.

There are drawbacks to contact breakers in high performance work, however. The standard contact breaker spring will weaken if subjected to very high engine speeds, beyond about 6000 rev/min. And within the 5000 to 6000 rev/min speed range the problem of points bounce may begin to manifest itself. At worst, the cam follower will skip from lobe to lobe on the cam, barely allowing the points to close, with the result that there is insufficient time for the magnetic field to build up in the coil. This will produce misfiring and loss of power.

High performance distributors usually have contact breakers with stronger than standard springs to extend the usable engine speed range before the onset of points bounce. It is sometimes possible to fit this type of contact breaker into a standard distributor with very little modification. But this will only be necessary on an engine which is frequently required to run at high speeds.

Other disadvantages of conventional contact breaker ignition are discussed in **Section 3:5**. Some V8 engined cars have twin contact breaker sets and a four lobed cam. This arrangement to some extent overcomes the problem of the very short coil recovery times allowed by an eight lobed cam and reduces the possibility of points bounce.

Dwell angle:

The width of the gap between the two points determines a factor called the dwell angle. Simply, this is the number of degrees of cam rotation that occur while the contact breaker is closed. The wider the contact breaker gap the longer time the points take to close and the smaller the dwell angle – conversely the narrower the gap the greater the dwell angle. Two factors are involved here. Firstly, the coil needs at least 2 milliseconds to recover its magnetic field sufficiently to deliver another impulse; high speeds and a wide gap (which decreases the dwell angle) may mean the coil recovery time falls below this minimum. Secondly, the wider the gap the more the effective ignition timing is advanced, the narrower the gap the more the timing is retarded.

This means in practice that the contact breaker gap is one of the most important settings on the car, having wide ranging effects on both economy and performance.

Advance mechanisms:

The base plate to which the contact breakers are secured is in two parts and designed in such a way that the points have a small degree of freedom to rotate around the cam. This movement is controlled by a vacuum diaphragm acting against a return spring. The diaphragm is a means of adjusting the timing to engine load as outlined above (see **FIG 3:4**).

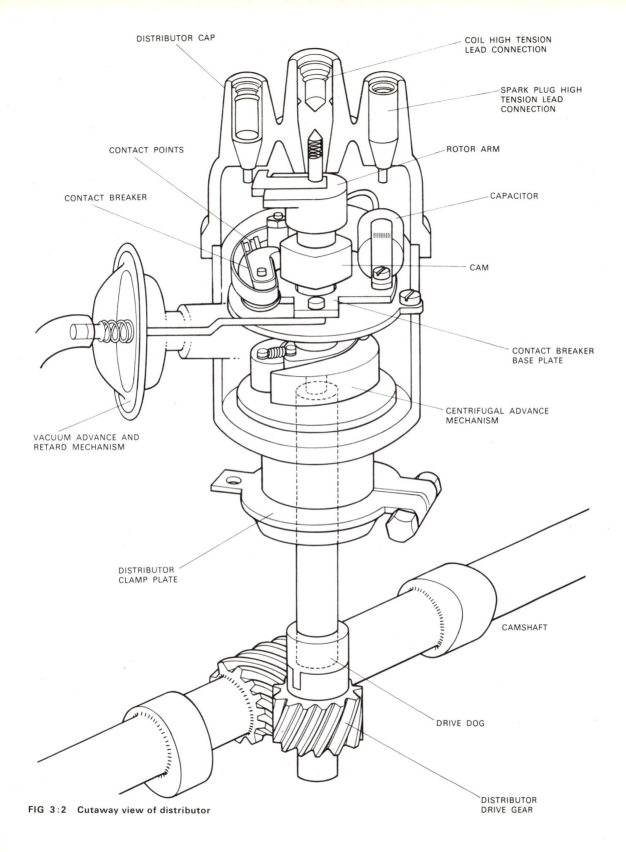

DISTRIBUTOR CAP

COIL HIGH TENSION
LEAD CONNECTION

SPARK PLUG HIGH
TENSION LEAD
CONNECTION

CONTACT POINTS

ROTOR ARM

CONTACT BREAKER

CAPACITOR

CAM

CONTACT BREAKER
BASE PLATE

CENTRIFUGAL ADVANCE
MECHANISM

VACUUM ADVANCE AND
RETARD MECHANISM

CAMSHAFT

DISTRIBUTOR
CLAMP PLATE

DRIVE DOG

DISTRIBUTOR
DRIVE GEAR

FIG 3:2   Cutaway view of distributor

51

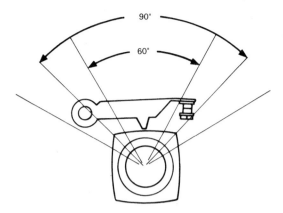

FIG 3:3 Contact breaker dwell angle

The depression or vacuum created in the engine's induction manifold is applied to a thin bore plastics or metal pipe connected between the carburetter and the vacuum diaphragm unit. By this means the base plate mounted contact breaker, moving in sympathy with the induction state of the engine, is made to switch the primary current off (and thus create the spark) earlier in the combustion cycle (advanced ignition) or later (retarded ignition).

The cam's rotation about the drive shaft is part of the mechanism by which the ignition is advanced with engine speed. The automatic or centrifugal advance mechanism consists of two specially shaped weights pivoted in such a way that as they move outwards under the influence of the distributor shaft's rotation they turn the hollow cam around the shaft. The effect is to advance the ignition at speed. The cam is returned to its static

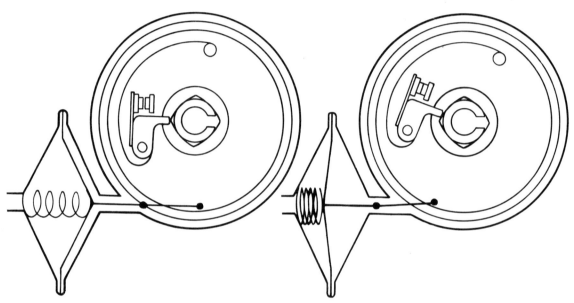

FIG 3:4 Operation of vacuum timing control

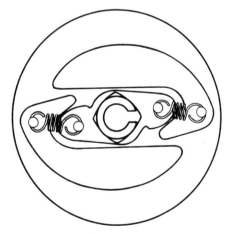

FIG 3:5 Operation of centrifugal timing control

position by two small springs designed to progressively control the action of the weights and closely match the advance to the engine's spark requirements (see **FIG 3 : 5**).

For tuning purposes there are usually two means of advance and retard adjustment. The simplest is the rotation of the whole distributor body in the clamp which secures it to the engine. This is used for basic adjustment during tuning and on some distributors it is the only means of adjustment. Usually there is also a vernier adjustment with a calibrated thumbwheel for fine tuning. This mechanism is often coupled with the vacuum advance.

The distributor capacitor:

One more very important component is housed within the distributor body or attached outside it – the capacitor or condenser. This is connected with its single lead to the moving contact breaker terminal. Its second connection is provided via the metal capacitor body to the base plate (earth). The capacitor performs two functions.

When the magnetic field collapses in the coil the lines of force intersect the primary winding as well as the secondary winding. The result is an induced current of up to 300 volts in the primary winding. This is called a back emf (electromotive force) because it flows in the opposite direction to the primary winding supply current. This pulse attempts to arc across the contact breaker points as they are opening but it is intercepted by the capacitor which soaks it up or suppresses it. Without the capacitor the arcing would rapidly burn and pit the contacts causing misfiring of the engine.

The capacitor momentarily stores the energy of the back emf before it feeds it back into the primary winding to aid the collapse of the magnetic field. This reinforces the high tension impulse from the secondary winding.

The ignition coil:

The process of magnetic field collapse that results in a high tension impulse has already been described. It is a simple process but just as the condition of the contact breaker and other distributor factors affect the spark intensity, so does the coil design.

The coil is a compact unit fitted inside a cylindrical aluminium can. The can contains two coils – the primary and secondary windings – packaged in an iron jacket and surrounded by a high voltage insulating material which also serves to dissipate heat (sometimes a special oil is used). The two windings – there are a few hundred turns in the primary winding and several thousand in the secondary winding – are formed concentrically (primary outermost) round a laminated core.

Coils are designed to produce a much higher voltage that that actually required for a spark. Plugs in good condition in an engine that is perfectly in tune may spark at 5000 to 9000 volts. Conditions are rarely this perfect and the usual minimum considered necessary is between 12,000 and 15,000 volts; but the standard coil will produce around 30,000 volts.

There are obstacles in the way of the high tension current like the air gap in the distributor cap and all these tend to reduce the voltage available at the spark plug. But by far the largest restriction on coil output is the car's system voltage – if it is lower than 12 volts, coil output suffers. In fact when the starter motor is turning the

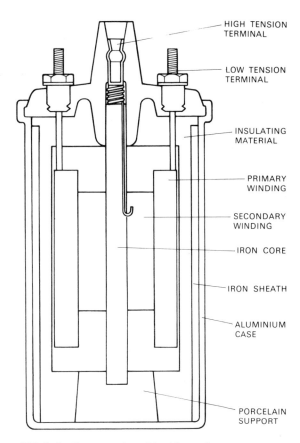

FIG 3:6   Cross-section of ignition coil

Labels on figure:
- HIGH TENSION TERMINAL
- LOW TENSION TERMINAL
- INSULATING MATERIAL
- PRIMARY WINDING
- SECONDARY WINDING
- IRON CORE
- IRON SHEATH
- ALUMINIUM CASE
- PORCELAIN SUPPORT

voltage available to the coil is in the region of 9.5 volts; with a partially flattened battery it may drop to 7 volts.

The hefty margin of output from the coil is enough to cope with these conditions and produce a high enough voltage to ignite the mixture at the spark plug.

An important part of tuning is to ensure that as high a voltage as possible reaches the plugs. Some of the operations are a matter of routine maintenance but other action can be taken, like fitting a high output 'performance' coil.

A coil of this type is designed to give a higher voltage secondary winding output. In practice this means that when the battery voltage is low during use of the starter motor a high enough voltage to fire a spark is still produced. At high engine revolutions, when the contact breakers may be opening 200 times a second or more and coil recovery time becomes very dependent on the correct adjustment of the dwell angle, higher voltages than normal will still be available.

However the most important requirement for the high performance coil is to provide the higher firing voltages necessary in an engine with a higher than normal compression ratio. The spark voltage required rises in almost direct relationship to the pressure of the mixture.

The availability of a higher voltage may also mean that a wider plug gap can be used. Plug gapping is a critical adjustment and should not be tampered with lightly. However in certain circumstances it is permissible to open out the gap a little (say .005 inch on a .025 inch plug gap) to increase the amount of mixture heated to above ignition point by the spark. The wider the plug gap the higher the voltage required for a spark and so the more the necessity for a performance or sports coil.

Ballast resistor ignition systems:

Modern standard ignition coils and many high performance units are often of the ballast resisted type. In this system a resistor is connected in series between the ignition switch and the coil. The coil is of a special design operating effectively at 6 to 8 volts instead of the normal 12 volts. The resistor is the means by which the car's system voltage is lowered to operate the coil when the engine is running.

During starting when the voltage available from the battery drops to as low as 7 volts (although normally 9.5 volts is the minimum) this reduced supply bypasses the resistor and is fed direct to the coil from the starter solenoid.

The advantage of this system is that the coil is always working at or near its normal primary winding voltage and peak voltage is produced from the secondary winding during starting.

The ballast resistor may be in the form of a resistance wire bound into the car's loom, or a small ceramic mounted unit fixed on or near the coil.

Spark plugs:

The spark plug itself appears to be a very simple item but in fact it is a highly developed component designed to withstand electrical pressures of up to 30,000 volts while operating with one tip in a combustion chamber at up to 900 deg. C and the outer end at freezing point.

The spark plug consists of a metal shell with a threaded shank that screws into the cylinder head. A ceramic insulator is fixed inside the shell with a special cement which also seals the spark plug to prevent gas leakage from the cylinder. Ribs on the insulator are intended to prevent flashover of the high tension current along the spark plug body. Running from the nose of the plug to the terminal top is the centre electrode and conducting rod. The spark plug gap is formed between the electrode tip and the earth electrode (see **FIG 3:7**).

One spark plug may look just like any other – in fact there are numerous complex designs. Most of the changes centre around two factors; **1** the temperature of the combustion process; **2** the clearance allowed for protrusion of the plug into the combustion chamber.

In a high performance engine the heat generated in the combustion process is higher than that generated in a standard engine. The spark plug tip and side electrode are particularly vulnerable to heat. Firstly they may physically burn away and secondly they may glow sufficiently hot to cause ignition without a spark. The result may be damaging pre-ignition. Conversely the plug has to run fairly hot to clean itself of oil and carbon deposits that may coat the plug under certain engine conditions – optimum plug tip temperature is 500 to 600 deg. C.

Naturally combustion temperature characteristics vary from engine to engine and the answer has been to design ranges of plugs suitable for operation in varying temperature conditions. Plugs are designed to have differing abilities to conduct away heat from the tip of the centre electrode. 'Colder' plugs, those with enhanced heat dispersal properties, have shorter ceramic 'noses' and shorter lengths of centre electrode alloy. This configuration means that the point at which the heat is dispersed into the shank of the plug and thereby into the block is very much nearer the plug tip. 'Hotter' plugs, paradoxically for colder engines (low compression, low speed), have longer centre electrodes and noses (see **FIG 3:9**).

Problems of clearance between the piston at tdc and the tip of the plug and those arising from certain combustion chamber design characteristics have led to some variations in electrode shape and position.

The centre electrode of the conventional plug barely emerges from within the plug shank. Its sturdy side electrode may cover the nose or be slightly cut back to the centre of the nose.

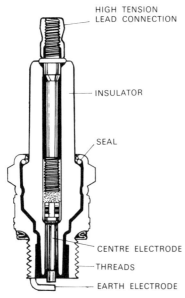

FIG 3:7   Cross-section of sparking plug

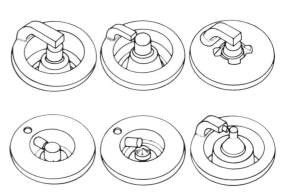

FIG 3:8   Types of plug electrode

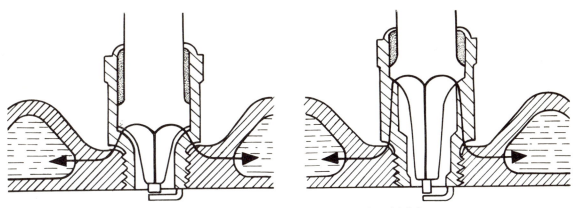

FIG 3:9 Cold plugs have shorter ceramic insulator noses (left) than hot plugs (right)

Recently car manufacturers with plenty of space in the combustion chamber have preferred to fit plugs with a slightly more protrusive centre and side electrode design. These have the advantage that incoming fuel assists in the cooling of the tip with the result that at lower engine speeds the plug runs hotter.

Where there is little or no clearance above the piston at tdc (for example where a raised crown piston is used)

or where valve clearance problems arise a fully retracted gap plug design may be used. In this type the centre and side electrode are fully shrouded inside the threaded shank. Generally speaking this design is only used for very highly tuned engines with superchargers or those burning alternative fuels (dragsters).

The choice of plug for a particular engine is extremely critical. It is rarely advisable to vary from that recom-

Spark plugs – heat range comparisons:

Plug size	AC	Autolite Motorcraft	Bosch	Champion	Lodge	KLG	NGK
14 mm $\frac{3}{4}$ inch reach standard design	44N C42N C41N 43XL S41XL	AG3 AG2 AG901 AG701 AG701 AG501	W175T2 W225T2 W260T2	N5 N4 N3 N2 N60 N57	HLN 2HLN 3HLN 3HLN	FE70 FE80 FF220 FE100	B6ES B7ES B9ES
14 mm $\frac{3}{4}$ inch reach protruding electrode	C42N	AG32 AG22 AG12	W200T30 W260T28	N9Y N66Y N65Y N64Y N63Y N60Y	HLNY 2HLNY 2HLNY 3HLNY 4HLNY 5HLNY	FE65P FE125P FE135P FE155P	BP7E
14 mm $\frac{1}{2}$ inch reach protruding electrode		AE32 AE22	W175T1 W175T7 W225T7	L87Y UL82Y L66Y L64Y L61Y	CNY	F55P F65P	
14 mm tapered seat protruding electrode	44TS 441TS 40TS 436TS	AF22 AF12	WA200T40	BL9Y BL7Y BL64Y BL60Y BL57Y			BP7FS
18 mm tapered seat protruding electrode	83TS	BF32 BF22 BF12 BF601		F9Y F7Y F62Y F60Y	HTNY	MT65T	A7F

mended by the car manufacturer as far as the temperature range is concerned. Where the compression ratio has been raised, the timing advanced from standard and the mixture deliberately weakened it will be necessary to experiment with plugs at least one grade colder. But for normal gas-flowed, mildly tuned and economy cars no change of temperature grade is recommended.

However that is not to say that experimentation with the equivalent plugs of various manufacturers cannot pay dividends in performance. Minute differences in temperature range, configuration of electrodes, and so on, can make surprising differences to the engine's power output. In the accompanying table some approximate equivalent grades of plug are given. In each size or type range the top of the list represents the hottest plug in the range.

While juggling with the equivalent types of plug from various manufacturers is an excellent measure to gain performance practically the only way to assess the gains is to carry out road or dynamometer testing, substituting each set of plugs in turn. As there are seven major manufacturers and one or two smaller makes this could be very expensive. It is not therefore recommended that this be a standard part of economy tuning. One way to get round this is to try out a different set of plugs every time a change is necessary; or it may be possible to borrow a set of a different type from a friend or colleague for a fuel consumption trial over a regular reproducable commuter run (see **Chapter 10**).

Polarity warning:

It is absolutey essential that the ignition coil is connected so that the polarity of the high tension at the centre electrode of the plug is negative – regardless of the earth polarity of the car's system. This is because the spark is thus made to travel from the centre electrode outward to the side electrode, a critical combustion factor at high speed. Ignoring this warning can result in high speed misfire, prematurely burnt out plugs and potential damage to the piston crown.

In practical terms, this means that the polarity of the coil low tension connections must be correct. Confusion is unlikely if the coil terminals are marked + and –. On a negative earth car the terminal marked – must be connected to the contact breaker. But some coils are marked CB (for contact breaker) and SW (for switch) or have numbers corresponding with those in the car's wiring diagram; these markings are for a system of specific polarity and an error can arise if such a coil is fitted to a car other than the one for which it was intended. If in doubt fit a new coil, as the design of some coils renders them less efficient, even if the polarity is correct, when the connections are the reverse of those intended.

3:2 Practical ignition tuning

Ignition coil, spark plugs and leads:

The most important tuning to be carried out on the high tension side of the ignition system is to ensure that it is maintained in good condition and that means paying close attention to the firmness of HT connections, the continuity of the lead insulation, cleanliness and freedom from moisture.

Wipe away grease and dirt, which could retain moisture, from around the coil body. Check the coil can is firmly secured in its mounting. Ensure the insulating shrouds are in good condition and the push fit connections are firmly made.

Two main types of high tension lead are in use on today's cars; one has rayon fibres impregnated with carbon through the core and the other has stranded copper wire. Both have thick plastics insulation – older leads may have a synthetic rubber insulation. Replace old leads and inspect leads regularly for minute cracking and crazing. The surface of the wires should be kept as clean and dry as possible. Label the leads clearly to distinguish which cylinder they serve. Leads can be cleaned with a methylated spirit soaked rag.

During the winter especially, spray HT leads, distributor cap, ignition coil and all the shrouds with a water repellent aerosol preparation. Two types are commercially available – one is the multi-purpose lubricant and repellent often called WD40 after the American government standard to which it is made and the other is a quick drying lacquer. There are several brands of each type on the market.

For economy the minimum service interval for spark plugs should be about 3000 miles – plugs should be replaced between 9000 – 10,000 miles.

Always use a good quality plug spanner with a short lever to remove plugs – take every care to ensure that dirt does not enter the plug hole during removal. Take special care when removing plugs from an alloy head, it is very easy to strip the threads. Stripped threads can be repaired using a specially designed helical steel coil insert, a job best carried out by experts.

Before cleaning the plugs, examine them for clues about the engine's condition and state of tune.

Light tan or grey deposits with electrode wear of up to .001 inch per 1000 miles indicate good engine and ignition condition.

Dry sooty black deposits on all plugs indicate either weak ignition, retarded timing, low compression or, most usually a rich mixture – too rich for economy.

Overheated plugs with white tip deposits and accelerated electrode wear point to over-advanced timing, worn distributor or a weak mixture (this may mean manifold air leaks).

Short plug life (accompanied by high speed misfire) and a characteristic dished pattern of wear on the earth electrode mean the polarity of the coil is reversed.

One or two plugs with burnt or melted electrodes indicate pre-ignition in the cylinder or cylinders concerned – this is perhaps as a result of glowing combustion chamber or plug deposits, or a plug that is too hot for that compression ratio or mixture. But this uneven distribution of symptoms can indicate a cam or distributor wear problem that is only affecting the cylinders in question.

Examine the porcelain insulator of the plug for cracking – discard the plug if any is found as it renders the insulation ineffective. Plugs are best cleaned on a garage grit blaster but a second best method is to clean the threads, plug tip and earth electrode with a fine wire brush. After cleaning use a contact file to square off the end of the centre electrode and the part of the side electrode opposite the central tip.

Reset the plug gap to the car manufacturer's recommended dimension – usually between .023 inch and .030 inch – with feeler gauges. Use a special spark plug bending tool for this. Always bend the side electrode – never attempt to bend the centre electrode.

Most plugs have a special metal gasket to ensure that the joint with the cylinder head is gastight. Before replacing a plug make sure that this is not completely flat or broken. (Some plugs have a tapered seat – see that this is in good condition.) Plugs should· be replaced finger tight (a smear of graphite grease helps especially on alloy heads but do not use any other lubricant) and then tightened with a quarter turn more of the plug spanner. Clean the plug's ceramic insulator with a methylated spirit soaked rag.

The distributor:

Most of the operations described here are part of normal maintenance. However they have such a bearing on the car's performance and economy that they must be considered as an integral part of the tuning process. The sequence described is that for a Lucas distributor – major differences between these and other manufacturer's units are detailed at the end of the section.

1 Distributor cap and rotor arm. The distributor cap is removed by levering off the two steel clips hinged to the side of the unit. Remove the cap with plug leads. Examine the cap for signs of tracking (jagged streaks of dirt between the contacts inside the cap usually give away this high voltage short circuit) and look out for hairline cracks – renew the cap if either fault is found. If the cap is sound clean out the inside with a dry cloth.

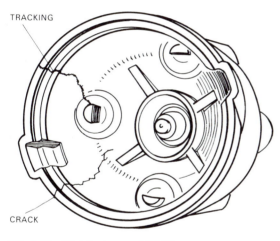

TRACKING

CRACK

FIG 3:10 Distributor cap defects

Examine the condition of the carbon brush in the centre of the cap. A fixed brush worn flush with its mounting means renewing the entire cap – spring loaded brushes that are worn, or have corroded springs can simply be renewed.

Carbon deposits on the cap electrodes can be cleaned off with methylated spirits (do not use abrasives as the rotor arm gap will be widened) but pitted and burnt electrodes mean cap renewal.

Pull or carefully lever the rotor arm off the end of the distributor shaft. Examine it for cracks or tracking, renewing it if necessary. Clean away any dirt deposits with methylated spirits; but renew the arm if it is pitted or burnt.

2 Contact breaker points. Open the contact breaker points by rotating the engine until the cam lobe lifts the contact breaker heel. To turn the engine, use a spanner on the engine crankshaft pulley nut, pull the fan belt or rock the car to and fro in top gear. Inspect the contact faces for pitting, burning or other damage – renew the contact breaker if any fault is found. Dirty points can be cleaned with a cloth.

For maximum economy and performance contact breakers should be replaced at a maximum of 6000 miles.

Remove the old set, carefully noting the order of screws, washers and tag connectors, by undoing the nut securing the moving contact spring, extracting the plastics 'top hat' insulator, pulling aside the low tension leads and taking the moving contact off its pivot. Take the insulating washer off the pivot and undo the fixed contact securing screw – the fixed contact can then be removed.

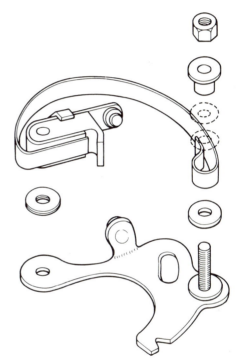

FIG 3:11 Lucas two-piece contact set

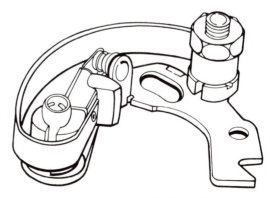

FIG 3:12 Lucas one-piece Quikafit contact set

The new contact breaker set should be of the correct type for the particular distributor. Two types of set are available – one-piece or two-piece. For many Lucas distributor applications they are interchangeable.

To refit a two-piece set locate the fixed contact on the main pivot pin and replace the securing screw but do not fully tighten it. Put the insulating washer on the spring pivot, followed by the moving contact spring. Reposition the leads and insert the 'top hat' insulator through them and over the pivot. Tighten the nut down on the insulator.

A one-piece set is simply located on the main pivot pin and secured by the fixing screw. The leads are secured to the contact by a plastics nut which must not be overtightened.

3 Contact breaker gap adjustment. Check that the moving contact heel is on the highest point of the cam so the points are fully open. Place the blade of a screwdriver in the adjustment slot and twist the screwdriver to fully open the points. Insert a feeler gauge (.015 inch) between the points and, turning the screwdriver, carefully move the fixed contact plate to close the points on to the feeler gauge. The feeler should just be pinched by the points with almost no tension placed on the moving contact. Tighten the fixed contact screw, move the cam through one revolution and recheck the gap, adjusting if necessary. Lucas allow an adjustment range of .014 inch to .016 inch for all applications of their distributors. Contact breakers 'bed in' as the follower conforms to the cam. It is wise to recheck the gapping after 500 miles. For race and rally cars use only contact breaker sets that have been bedded in.

Some point sets are protected by a waxy film over the contacts. This must be removed by wiping with a petrol soaked rag. Ensure that on reassembly the low tension leads are positioned away from the distributor's moving parts.

4 Cam and contact breaker lubrication. Inspect the cam faces for scoring and corrosion patches – renew the cam if there is evidence of either. The cam faces should be lightly smeared with high melting point grease (such as Retinax A). Two drops of oil should be applied to the cam screw and another two drops should be trickled into the gap between the base plate and cam. A drop of oil on the pivot of the moving contact breaker is sufficient – use a tiny smear of grease on the pivot of a one-piece set. Carry out lubrication carefully and clean off any excess grease or oil. Stray lubricant inside the cap or on the contact breaker points can cause misfiring.

5 Checking vacuum advance. The vacuum advance diaphragm unit is located on the side of the distributor body – a plastics or metal pipe connects it to the induction manifold on or near the carburetter flange. Check the operation of the vacuum diaphragm by undoing the push fit connection at the carburetter and sucking hard on the end of the pipe maintaining the vacuum by placing the tongue over the end of the tube. The base plate should move and maintain its position until the vacuum is released.

If the base plate does not move, first check that it is free to move and that the vacuum tube is obstruction free (remove it completely and blow through it). If the tube is clear the fault lies in the diaphragm unit – it is either pierced or jammed. A slow return of the base plate while maintaining vacuum also means a pierced diaphragm.

Faulty diaphragm units can be removed by unhitching the operating spring from the base plate, removing the spring clip and fine tuning nut (do not lose the retaining spring), and withdrawing the unit. Retune the engine by static or stroboscopic means (see later) after fitting a new diaphragm unit. Many high performance distributors and some standard units do not have a vacuum advance unit as it would tend to over-advance the spark at high engine speeds.

6 Checking centrifugal advance mechanism. The weights of the centrifugal advance mechanism should be free to move outwards against the tension of the restraining springs. To check this movement replace the rotor arm on the distributor shaft, and determine the normal direction of rotation (it is often marked by arrow on the rotor arm). Grip the rotor arm firmly and rotate the cam in the direction of rotation. On releasing the rotor arm it should spring back if the weights are free to move. The movement should be positive and free from stickiness.

If they are not operating correctly, remove the distributor base plate (two or three screws locate it on the distributor body) to gain access to the weights. Note their mounting position before removing and cleaning them, paying particular attention to the pivot points. Replace the weights and fit new restraining springs.

Most modern distributors have two different springs to give a progressive matching of spark advance to engine requirements. It is most important to replace them with the correct springs and to fit them in the right positions. Using the wrong springs, stretched or damaged springs, or fitting them incorrectly can seriously impair the engine's efficiency.

7 Other maintenance checks. Ensure the capacitor earth connection to the base plate is tight and that the capacitor lead to the moving contact is routed away from moving parts. Inspect the wiring of the coil lead terminal (a Lucar spade fixed in a plastics insulator slotted into the distributor) for security and clean away dirt from around internal and external connections. Try side to side movement of the cam to check for any play between the distributor shaft and cam and between the shaft and its plain bush bearing. If play is visible it is best to fit an exchange unit. Distributor wear can cause erratic timing and consequently poor running.

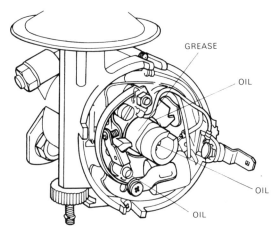

GREASE

OIL

OIL

OIL

FIG 3:13   Lubricating the distributor

Variations in distributor design:

AC Delco (Vauxhalls):

The centrifugal advance weights on some AC Delco units are mounted above the base plate and contact breaker points. The rotor arm is a circular plastics cap secured by two screws. The rotor arm's centre contact is a sprung leaf which must be bent up to ensure it contacts the fixed brush in the distributor cap (see **FIG 3:14**).

The contact set is similar to Lucas except that the low tension leads are secured by a plastics clip. The gap for most AC Delco points is .019 inch to .021 inch.

Motorcraft/Autolite (Fords):

The rotor arm has a sprung centre electrode contacting a fixed brush in the cap (see **FIG 3:15**).

A compact one-piece contact set is used and the low tension leads are secured by a terminal screw. Vacuum advance is best checked by revving the engine and watching for movement of the external connecting link between the diaphragm unit and the distributor body.

Bosch (some German, Swedish and Italian cars):

The only major difference between Bosch unit and Lucas is that a circular condensation shield is often fitted on the distributor shaft under the rotor arm. The moving contact is retained on its pivot by a small spring clip – take care not to lose it in the distributor body during removal. Points gap for most Bosch applications is .015 inch to .017 inch (see **FIG 3:16**).

Marelli (some Italian cars):

Two screws retain a circular rotor arm on the centrifugal advance weight mounting which is above the contact breaker and base plate. Some older Marelli units have an external wick for lubrication of the shaft. Points gap is between .018 inch and .020 inch (see **FIG 3:17**).

Ducellier (some French cars):

A two-piece contact set is attached to the base plate by a screw and spring clip. A special tool is recommended for points setting (.017 inch to .020 inch), but the job can be done with a screwdriver. There is a felt pad at the centre of the cam for lubrication (see **FIG 3:18**).

Hitachi (Datsun):

Conventional distributor with a one-piece contact set, weights fitted beneath the base plate, and the capacitor mounted on the outside of some units. Points gap usually between .018 inch and .020 inch (see **FIG 3:19**).

3:3 Static ignition timing

Once set, the ignition timing (simply, the point in the combustion cycle at which the distributor delivers the spark impulse to the plug) should not vary provided that the base plate is secure inside the distributor and the distributor is firmly clamped onto the engine. However well the marks have been lined up it is wise to retime the ignition after removing and refitting the distributor – the

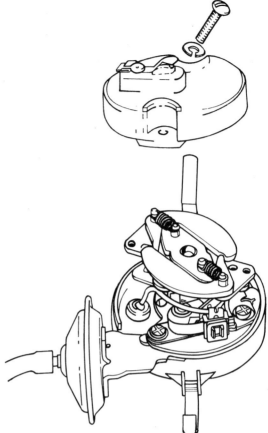

FIG 3:14   AC Delco distributor

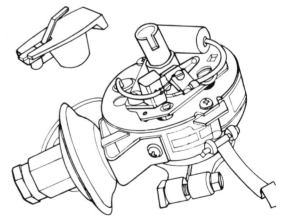

FIG 3:15   Motorcraft/Autolite distributor

procedure is also necessary when a new distributor is fitted and when new contact breaker points are fitted.

There are two ways of setting the ignition timing, the static and stroboscopic methods. A few modern cars, notably Audi, some VWs, Datsun and Volvo, should only be timed with a stroboscope although the static timing

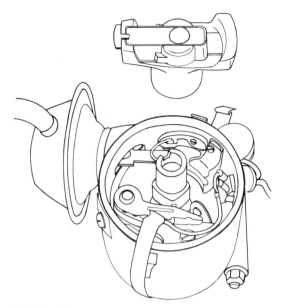

FIG 3:16　Bosch distributor

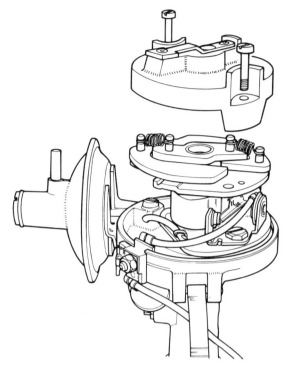

FIG 3:17　Marelli distributor

procedure can be used during engine assembly to achieve an approximate distributor position.

All engines have timing marks – one or more marks on the crankshaft pulley, or, in many transverse engined cars, on the flywheel, which align with a fixed mark or marks on the cylinder block or flywheel housing. When the marks are on the flywheel there is an inspection hole in the flywheel/clutch housing covered by a small plate. Very often the only way to see the marks through the hole is to use a mirror and a torch. Keep the mirror in position by sticking it to the flywheel housing with a lump of putty or Plasticene.

In most cases there is a pair of marks which align when the piston in the timing cylinder – usually No. 1 cylinder, sometimes No. 4 or 6 (refer to the car handbook) – is at the top of its stroke, a position referred to as top dead centre or tdc.

For most engines the correct timing point is a certain time, measured in degrees of crankshaft rotation, before top dead centre, btdc – possibly up to 15 deg., the precise figure being given in the handbook or servicing instructions for each model. With the advent of emission control regulations some models for some markets are timed at or even after top dead centre (atdc).

Often a timing scale is provided in addition to the tdc marks so that the correct timing point can easily be found. With a pulley turning clockwise (most engines) fixed marks for points btdc will be anticlockwise of the tdc mark, marks on the pulley for points btdc will be clockwise of the tdc mark.

A few exceptions, like the later VWs already mentioned, have only a single pair of marks indicating not tdc but the actual timing point.

If a precise timing mark is not provided cut a circle of card the same diameter as the crankshaft pulley. Draw a line on the card from the centre to the edge. Using a protractor draw a second line from the centre to the edge the correct number of degrees btdc (or atdc) from the

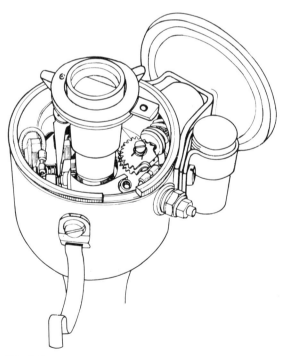

FIG 3:18　Ducellier distributor

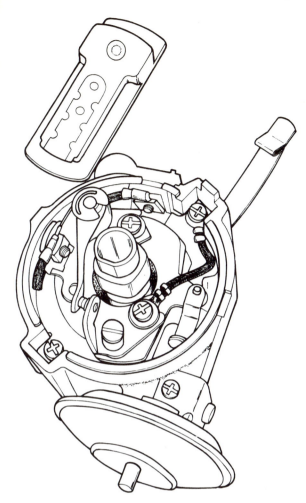

FIG 3:19 Hitachi distributor

first line (see **FIG 3:20**). Place the card over the pulley with the first line on the pulley's tdc mark. Using paint or a punch or scriber, make a timing mark on the pulley at the point indicated by the second line.

Use the following procedure for static timing.

1 Ensure the timing cylinder is near top dead centre on its compression stroke; this may be achieved by rotating the engine using a spanner on the crankshaft pulley nut until a finger or thumb held tight on the plug hole of the timing cylinder is forced off by the pressure. Remove the distributor cap and check the rotor arm is adjacent to the electrode for the timing cylinder.

2 As accurately as possible, line up the timing marks at the correct number of degrees btdc (or atdc).

3 With the ignition on, connect a test lamp between the low tension contact on the side of the distributor and a good earth.

4 Slacken the distributor clamp bolt until the body of the distributor can be turned freely.

5 Ensuring that the rotor arm remains in the correct position to fire the timing cylinder, rotate the distributor body until the test lamp lights up, indicating that the

points have opened. Try this adjustment a few times until the exact point at which the light comes on can be determined. Tighten the distributor clamp bolt without disturbing the position of the distributor.

6 Check the accuracy of adjustment by slowly rotating the engine for a few turns. The lamp must light at the point when the timing mark lines up with the reference point. On some distributors fine adjustment can be made using the vernier screw. Repeat the procedure if any inaccuracy is found in the test.

An alternative method sometimes recommended involves connecting the test lamp across the two low tension contacts on the ignition coil. In this case the lamp will go out as the points open, so the distributor must be set at the exact point at which the light is extinguished. To avoid confusion decide which method is easier to use on your car and stick to it.

3:4 Stroboscopic ignition timing

A stroboscopic timing light (strobe) is used for this timing method which is performed while the engine is running. For really accurate results it is also necessary to have a reliable tachometer (rev counter). When the specified revolutions for stroboscopic timing (see car handbook) are in the region of 600-750 rev/min reasonably accurate results can be obtained by setting the carburetter idle adjustment to a slow, smooth idle.

There are two types of strobe used for timing – the cheapest and simplest is connected in series with the plug lead of the timing cylinder. A more expensive type, often with an adjuster and dial to set the number of degrees required, has a sensor which is fitted in series with the timing cylinder plug lead and low tension leads which are clipped to a live battery supply point (such as the live side of the starter solenoid) and earth (the car body).

Use the following procedure for stroboscopic timing.

1 Check that the engine's timing marks are clearly visible – it is useful to pick them out in white paint or chalk.

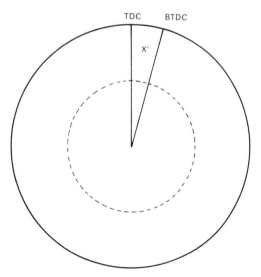

FIG 3:20 Timing card

2 If the handbook indicates the timing should be carried out with the vacuum advance mechanism disconnected remove the vacuum pipe at the carburetter end. Plug the carburetter vacuum connection – a short length of rubber or plastic pipe with a small lightly greased bolt stuck in the end is a simple solution to this.

3 Connect up the timing light and start the engine. Adjust the carburetter idle screw so the engine runs at the required speed. (A few cars are timed at quite high revolutions – up to 3000 rev/min – and for these it may be necessary to have a helper to operate the accelerator).

4 Point the light at the timing marks. Adjust the fine tuning vernier of the distributor until the correct degree timing mark appears stationary in line with the reference mark. If there is no vernier adjustment or the timing is so far out of true that the fine adjuster cannot bring it into line, get a helper to loosen the distributor clamp bolt and turn the distributor until the marks come into line. Extreme care must be taken during this procedure to ensure that the distributor remains stationary while the bolt is re-tightened.

The strobe light can also be used to check the operation of the centrifugal advance weights and the vacuum advance mechanism. With the vacuum disconnected, blip the throttle to raise engine revs and the timing marks should appear to move steadily backwards, returning to the correct timing when the engine slows. This procedure shows the centrifugal advance is working. Reconnect the vacuum pipe, rev the engine to a steady 2000 rev/min and disconnect the pipe. The timing marks will appear to move steadily forward on disconnection

of the pipe and on reconnection the marks will appear to move backward. If this effect is not observed fit a new vacuum advance unit.

Use the timing card method described for static timing to mark the crankshaft pulley with the advance marks for the vacuum mechanism and the automatic advance to make an exact check of the operation of the two mechanisms. The correct figures can be found in a workshop manual for the car in question. This check is most important on high performance tuned engines to ensure that the ignition is not over-advanced at high speed. On front wheel drive cars take the opportunity to mark the advance points on the flywheel when changing a clutch.

3:5 Alternative ignition systems

The conventional ignition system described so far has a number of weaknesses which may not show in the day to day use of a well-maintained family saloon but can cause trouble on high performance engines and those with more than four or six cylinders. The requirements of these engines and more stringent laws on exhaust emissions have led to the development of more sophisticated solutions to spark distribution and timing. Some more advanced systems for ignition are available to the enthusiast and family motorist alike in kit form; others will become commonplace on the lower priced saloons of tomorrow.

By far the weakest link in the ignition system is the contact breaker. It carries a current of up to 5 amps causing arcing from point to point and metal transference

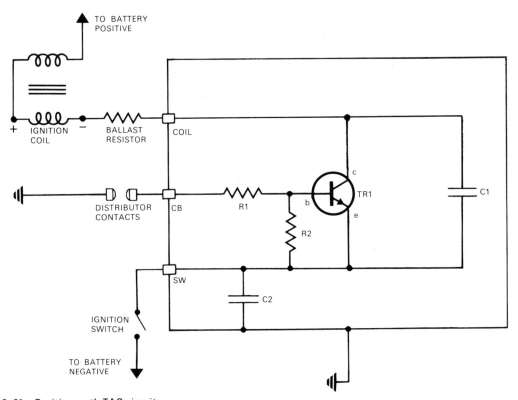

FIG 3:21 Positive earth TAC circuit

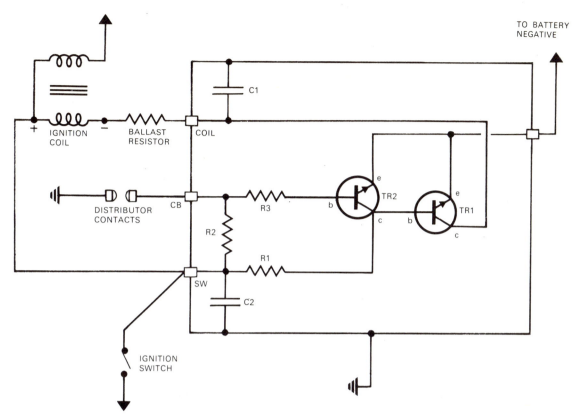

FIG 3:22   Negative earth TAC circuit

from one point to the other. Even if the points are properly gapped it is possible that uneven metal build up on one spot on a point will narrow the effective contact area and reduce voltage available to the primary winding. There are also the limitations caused by high speed points bounce.

Several systems which eliminate contact breakers are currently on the market. In addition, there are many systems which decrease the current load on the points, reducing metal transfer and wear.

Transistor assisted ignition:

Some of the problems associated with the use of mechanical contact breaker points can be overcome by performing the switching of the low tension current to the primary winding with a power transistor. In a transistor assisted ignition (sometimes called TAC – Transistor Assisted Contacts) system the points are relegated to switching the transistor control current of about 100 milliamps (instead of up to 5 amps) so there is no arcing across the gap and points wear is greatly reduced. The transistor has a very clean switching action, a definite break of the circuit, so the back emf from the coil cannot possibly maintain the primary current and certainly cannot arc. So a capacitor is no longer needed. On positive earth systems one transistor can fulfil the ignition switching but problems of polarity necessitate the use of two transistors for negative earth systems.

1 Operation of positive earth system. When the contacts are closed current flows in the transistor's base circuit and the transistor is conductive across the collector and emitter leads, thus the primary winding is energised (see **FIG 3:21**). When the contacts open the transistor becomes non-conductive and the coil field collapses to produce the high tension impulse. Resistors R1 and R2 reduce the current flowing through the contact breaker points. Capacitor C1 protects the transistor from high voltage surges and capacitor C2 is a radio interference suppressor.

2 Operation of negative earth system. Transistor TR2 in **FIG 3:22** is introduced into the circuit to control the switching action of TR1. With the contact breaker closed the base of TR2 is shorted to earth so no current can flow in its collector-emitter circuit. But current can flow through R1 to the base of TR1 which allows the primary current to energise the coil. When the points open TR2 becomes conductive making an easier current path than to the base of TR1 – so TR1 becomes non-conductive and the primary winding current is switched off.

The major advantage offered by TAC ignition is reduced contact maintenance. A points set will run for up to 25,000 miles without adjustment or replacement. As there is no variation between switching performance in hot or cold engine conditions a constant high voltage spark is obtainable from the coil. There is some truth in the claim that a system of this kind saves petrol – simply because it is possible to neglect distributor maintenance, as so many people do, without deterioration of ignition performance. It is not, however, a panacea for all ignition ills – TAC just maintains peak ignition conditions for longer.

Capacitor discharge ignition systems:

There are several kits on the market offering a form of ignition system based on capacitor discharge – well-known examples are the CD, Mobelec and Jermyn systems.

Capacitor discharge ignition systems combine the advantages of a transistor assisted circuit which relieves the contact breaker points of carrying the full primary current with a means of stepping up the voltage supplied to the coil primary winding. Inside the finned heat sink black box, the heart of a typical system, the battery voltage is transformed up to about 400 volts dc while the contacts are closed. The transformed current is stored in a capacitor until the contact breaker opens – it is then discharged through the primary winding.

The advantages claimed for the capacitor discharge system are an even, high voltage output from the coil secondary winding assisting cold-start conditions with the addition of electronic control of the dwell angle, eliminating the effects of points bounce.

These systems can also be combined with a method of eliminating the contact points altogether, similar in operation to the Lucas Opus system described later.

Capacitor discharge systems can be used with conventional ignition system coils but not with those fitted in ballast resistor systems.

Contactless ignition systems:

Most of the problems associated with conventional ignition systems can be eliminated by a combination of a ballast resistor coil and the replacement of the contact breaker by a transistorised (solid state) switching device. Kits offering this advanced principle are available – the best known are the Piranha and Lumenition types.

The Lumenition system (fitted as standard to some high performance cars) has a simple and elegant method for contact breaker replacement. The distributor cam is replaced by a segmented metal disc that is clipped over the now defunct cam in a conventional Lucas or Autolite distributor. Fixed to the base plate in place of the contact breaker set is a trigger unit consisting of a semi-conductor light source (gallium arsenide) with a lensed face arranged opposite a photo-transistor (a device that senses light). The main transistorised control unit provides current to the coil primary winding when the light beam from source to sensor is unbroken. But when a segment of the disc rotating on the distributor shaft passes between the light source and photo-transistor a signal passes to the control unit. The unit switches off the current to the coil primary winding causing a spark impulse in the conventional manner.

The advantage of this system is that once the position of the trigger unit is set on the base plate little attention needs to be given to the distributor apart from periodic lubrication and maintenance of the high tension circuit. The ballast resistor coil used with this system gives the normal advantages of good cold-starting performance.

The dwell angle is automatically fixed by the width of the disc's segments – the broader the segments the

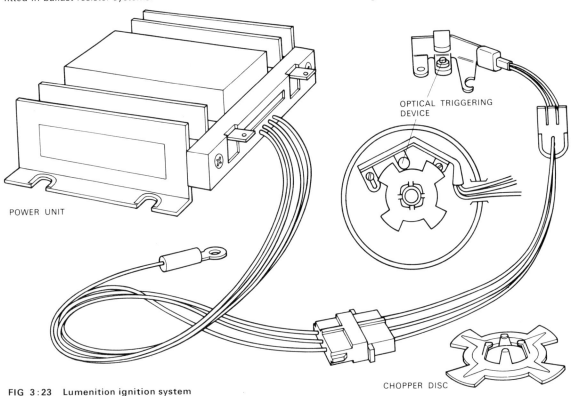

POWER UNIT

OPTICAL TRIGGERING DEVICE

CHOPPER DISC

FIG 3:23 Lumenition ignition system

longer the light beam is broken and the greater the dwell angle. This disc is manufactured to very close tolerances and ensures very precise timing.

The Piranha system varies only in detail from the Lumenition unit. It has a similar light source and photo-transistor trigger unit – but in this system the control unit switches off the primary winding current when a slot in the rotation discs allows the light beam to pass from lamp to sensor.

Lucas Opus ignition system:

Jaguar/Daimler V12 engined cars have an ignition system that is specially designed to cope with the ignition problems arising from the large number of cylinders (also to satisfy emission requirement in export markets). It is a development of the Lucas Opus system, once promised by Lucas as a kit conversion for conventional ignition systems but now exclusively used on these high performance cars. More export models and, possibly, some lower priced UK saloons may shortly be fitted with this system.

The Opus system consists of a special distributor, transistorised amplifier unit, ballast resistor and special Opus coil. The distributor has no contact breaker – switching of the amplifier unit is performed by small ferrite rod magnets mounted round the edge of a nylon disc rotating on the distributor shaft. The magnets, one per cylinder spaced at regular angular intervals, pass very close to a tiny magnetic pick-up mounted on the base plate (see **FIG 3:24**).

The pick-up is a transformer – as the magnets pass by a small current is induced in its primary winding, in turn inducing a current in the secondary winding. The current is a signal for the amplifier unit to switch off the coil primary current triggering the spark impulse in the conventional manner. The distributor also performs its normal function of passing high tension impulses on to the correct cylinder.

British Leyland all-electronic ignition system:

The ultimate ignition system which obviates the need for a distributor is now undergoing development by British Leyland and is expected to be fitted to higher priced saloons within the next two years – although probably on export models only.

It has been widely recognised that emissions problems caused by the mechanical aspects of conventional ignition systems (contact breaker maintenance requirements, wear and back lash in the distributor drive, and so on) can only be overcome by total electronic control of ignition. British Leyland propose a system in which the timing impulses are generated by a light interrupting blade on the clutch or flywheel (in direct contact with the crank-shaft) passing between a lamp and sensor mounted on the clutch/flywheel housing. A small integrated circuit computer, with up to 6000 transistor junctions, balances inputs from a vacuum transducer and the light sensor to advance and retard timing according to engine requirements. Coil generation of the high tension impulse is retained but an additional circuit controlled by a trans-ducer on the engine camshaft arranges the distribution sequence probably through relays in the high tension circuit.

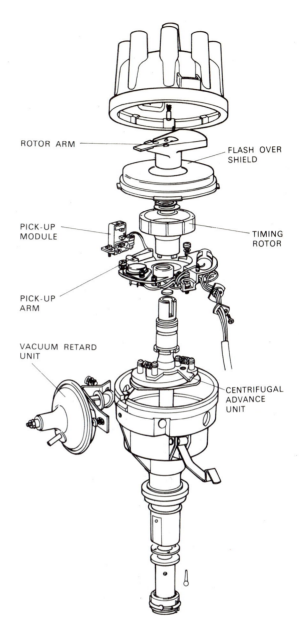

ROTOR ARM

FLASH OVER SHIELD

PICK-UP MODULE

TIMING ROTOR

PICK-UP ARM

VACUUM RETARD UNIT

CENTRIFUGAL ADVANCE UNIT

FIG 3:24 Exploded view of Lucas Opus distributor

3:6 Tuning summary – for economy

1 Maintain the ignition system in good clean condition – this ensures that the high voltage produced by the coil is not diminished by voltage leaks and tracking.
2 Replace components at the slightest sign of damage – high voltage will leak wherever the insulation is reduced especially where moisture gathers, for example, in minute cracks.
3 Replace contact breaker points at 5000-6000 mile intervals – replace plugs at 9000-10,000 miles.
4 Maintain the contact breaker gap as close as possible to

the specified setting. Check the gap every 2000-3000 miles and always check the gap 500 miles after fitting a new set of points.

5 Check the spark plug gap every 3000-4000 miles. In most cases maintain the gap at the specified setting. However it is permissible to experiment with gaps up to .005 inch wider than normal if the car's ignition system is in good condition or a sports (or high performance) coil is fitted. Investigate the fuel consumption by road-testing (see **Chapter 10**).

6 Generally speaking the ignition timing should be maintained at the specified setting – very slight advance or retard adjustments can be made in conjunction with road testing to establish the most economic setting for a particular engine.

7 Where a car's fuel octane requirement appears to fall between two of the British Standard star-rated grades, experiment with slight retarding of the ignition to check if it is possible to use the lower grade or alternatively always use a petrol station with a blender pump to obtain the exact grade required.

8 Ensure vacuum and automatic advance mechanisms are operating correctly. Replace any components showing signs of wear.

9 Regularly lubricate the cam spindle and smear fresh Retinax A grease on the cam when replacing the contact breaker points.

WARNING: Do not be lured by so called economy devices that are fitted into the high tension leads or at the plug cap. There have been several such devices on the market in the past and some may still be available. They do not work at all – various independent tests of such devices have shown that any effect they have is harmful rather than beneficial.

3:7 Tuning summary – for performance

Standard engines:

For engines with standard compression ratio, carburetter and manifolds.

1 Ensure all parts of the ignition system are in good condition, cleaned of dirt and grease, and free from moisture.

2 Replace contact breaker points and plugs at 5000-6000 miles and replace plugs at 9000-10,000 mile intervals.

3 Maintain the contact breaker gap as close as possible to the specified setting. Check the gap 500 miles after fitting a new set of points and every 2000-3000 miles.

4 Opening out the spark plug gap .004-.005 inch wider than the specified setting should be tried by road or dynamometer testing. Also test equivalent grades of plugs from various plug manufacturers – occasionally the

slight temperature range difference may prove advantageous to the performance of a particular engine.

5 It is not usual to find any performance advantage in varying the ignition timing on a more or less standard engine. Some minor variation may be beneficial to certain engines. However it is very easy to over-advance the timing in the quest for more performance. The result may be pinking or low-speed, full throttle detonation. But the greatest danger in over-advancing is high speed pre-ignition. This condition is usually inaudible and can cause a great deal of damage to the valves and pistons.

Modified engines:

For engines in which carburetter and manifold changes have been made or when the compression ratio has been raised.

6 Modified engines place a great load on the ignition system and almost certainly a high performance coil will be necessary.

7 High tension electricity of a higher voltage than standard requires particularly thorough attention to the condition of the system's insulation.

8 If particularly high rev/min are required from the engine the standard distributor will have to be changed for one with vacuum advance and automatic advance characteristics suited to the particular engine and application. It may be possible to find a distributor with the right characteristics from a high performance car in the same manufacturer's model range.

9 Contact breaker gap should be maintained at the standard setting but the possibility of fitting a stronger contact breaker spring should be examined – high performance distributors usually have points with a stronger spring. For consistently high revving engines transistorised or transistor assisted ignition is beneficial.

10 Ensure that the polarity of the high tension current is negative at the centre electrode.

11 The hotter the engine the colder the plug requirement. The temperature range of the plug is an extremely critical factor – consult plug manufacturers for advice on very highly tuned engines. A mild increase in compression ratio coupled with gas flowing, manifold and carburetter changes will normally only need a plug one grade cooler than standard. Experiment with the equivalent plugs of each manufacturer (see item **4**).

12 Timing should be carried out in conjunction with road or dynamometer testing. Considerable differences from the standard setting may be found to be beneficial. Observe the warning on the dangers of over-advancing given in item **5** of this section.

13 Pay particular attention to lubrication of the cam with hard grease like Retinax A, and light oiling of the automatic advance mechanism. Make regular checks for play in the distributor shaft bearings.

CHAPTER 4

Transmission

4:1 The transmission and tuning
4:2 Gearbox maintenance and overhaul
4:3 Modifying the gearbox
4:4 Clutch maintenance and overhaul

4:5 Modifying the clutch
4:6 The differential
4:7 Other transmission components

4:1 The transmission and tuning

The car's transmission is designed to transmit the power from the engine through to the road wheels. En route the power available is modified by the gearbox so the rate of work is precisely controlled to the driver's requirements. A low first gear gives snappy acceleration from rest, second gear allows a more progressive build up of speed which is continued into third gear while top gear might only just enable the engine to keep the car rolling at high speed. This is just one illustration of the use of various gear ratios in the gearbox. Completely different driving characteristics will be obtained by using a higher first gear. The move away from rest would need clutch slipping but as the engine revved more power would become available up to quite usable road speeds for town and traffic crawling.

In designing and building the gearbox the manufacturer of a mass production car has to make certain assumptions about the buyer's use of the vehicle and has to take into account the power available from a standard engine. The drivers of cars with tuned engines will almost certainly take a different viewpoint to the gearbox designer. The higher power available might make a very low first gear unnecessary or third gear may be too far into the last gasps of the engine's rev range at usable speeds. In short, transmission considerations can play just as great a part in the tuning of a car as beefing up the engine. It is not just the driver's average requirements that should be taken into account but also the engine tuner's modifications. A transmission built to deal with a standard engine may not be strong enough to cope with the new power available from a tuned engine.

Actual modification of the gearbox is certainly not on the economy tuner's list of cost-cutting procedures but the general principles set out in the following section on achieving a pleasant easy to use gear change entail little cost, make driving much easier and ultimately enable great savings to be made in servicing and repair. Altering the final drive ratio can have an effect on the car's economy and on some cars it may be possible to accomplish a considerable gain in miles per gallon for a reasonably small outlay by that means (see **Section 4:6**).

4:2 Gearbox maintenance and overhaul

The ease of gearshift operation is by and large an inherent characteristic of the gearbox design and the engine and transmission layout of the car. Front engine, rear wheel drive cars very often have a much easier shift action than transverse engined cars simply because the gearlever acts directly on the gearbox. Front wheel drive (fwd) cars usually have long remote control linkages which can impart a rubbery or notchy feel stemming from the linkage mountings. Particular manufacturers also seem better at gearbox manufacture and design than others.

However, these characteristics can be made much worse or sometimes improved by factors like the tolerances accepted in production and wear. But easier operation is usually obtainable only by rigorous attention to maintenance and a programme of modifications that overcome some of the shortcomings in gearboxes straight from the production line.

There is an enormous variety of gearbox designs. The points below summarise the main operations involved in regular and overhaul servicing of the unit. Consult the car's workshop manual for full details of removal, stripping down and repair.

1 Manufacturers quote varying time intervals for the topping up and replacement of transmission oil. Whatever

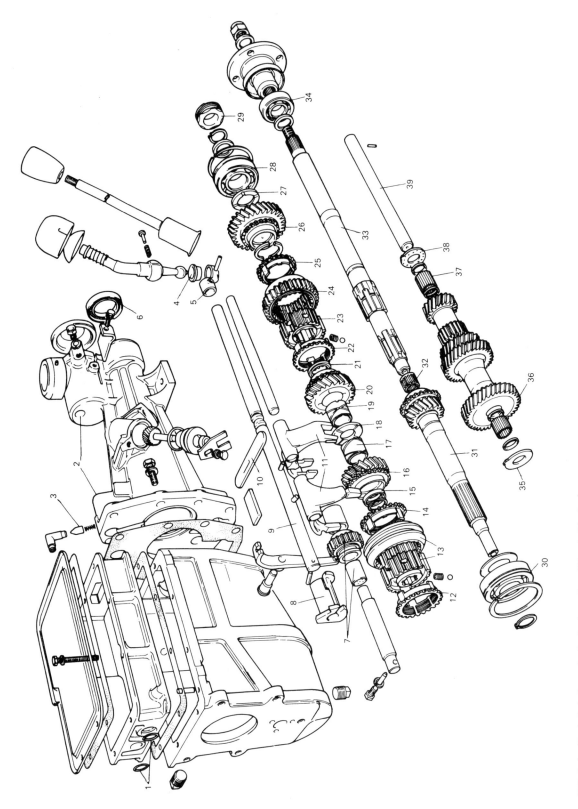

FIG 4:1   Exploded view of a typical gearbox (Morris Marina)

Key to Figs 4:1 and 4:2   Components most likely to show wear: 1 'O' ring 2 Selector shaft bearings 3 Detent plunger and spring 4 Lower gearlever seat 5 Gearlever yoke 6 Oil seal 7 Reverse idler gear and bush 8 Interlock spool 9 Gear selector shaft 10 Interlock spool plate 11 Gear selector forks 12 Synchromesh cup 13 Synchromesh hub and sleeve 14 Synchromesh cup 15 Thrust washer 16 Third-speed gear 17 Bush 18 Selective washer 19 Bush 20 Second-speed gear 21 Thrust washer 22 Synchromesh cup 23 Synchromesh hub 24 Reverse gear and synchromesh sleeve 25 Synchromesh cup 26 First-speed gear 27 Thrust washer 28 Mainshaft centre bearing 29 Speedometer wheel 30 Ballbearing 31 Imput shaft 32 Needle roller bearing 33 Mainshaft 34 Ballbearing 35 Front thrust washer 36 Laygear 37 Needle roller bearing 38 Rear thrust washer 39 Layshaft

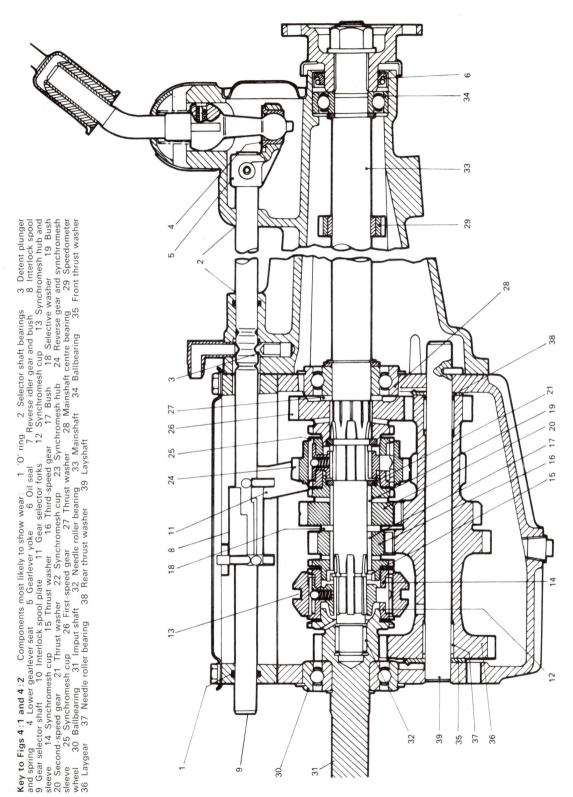

FIG 4:2   Sectional view of Morris Marina gearbox

69

the period stipulated for a specific car, a gearbox subjected to hard performance or competition use will benefit from more frequent changes of the specified lubricant. Some transverse engined cars have an integral engine and gearbox sharing the same lubrication system. Provided that the oil filtration system is working correctly and, where possible, a magnetic sump drain plug is installed, little improvement in operation can be gained by more frequent oil changes in this case. The only essential measure is to change the filter and oil at the correct interval.

2 Operation of the gearlever. The gearbox in some cars with front engines and rear wheel drive is positioned in such a way that it has been possible to mount the gearlever through the floor pan to give direct action on the selector mechanism. The lever is pivoted in a ball joint secured to the gearbox housing. Clean, grease and check the ball joint for wear at 6000 mile intervals. Nylon cup parts, often designed to cut down on vibration and noise in the gearshift mechanism, should be given especially close inspection as these can distort under heavy gearlever usage.

Front wheel drive and rear engined cars have remote control gearlever linkages. Play in this mechanism and the design of the mountings gives rise to the spongy or, perhaps, slightly slack feel in the gear change. Mechanisms enclosed in a housing, which also aids the stability of the bottom engine mountings, are often of the single rod type. The gearlever is mounted conventionally in a ball joint and its movement is transmitted by action through a fork or cup at one end of the rod. A similar fork or cup at the other end of the rod actuates a stubby lever at the gearbox end of the linkage.

Wear occurs at these two joints. In addition to checking that this does not affect the engagement of the gears the movement of the rod within the housing should be tested. Slight corrosion of the surface of the rod can interfere with smooth changes. Light oiling of the plain drilling or bush through which the rod runs will restore normal operation.

Some conventionally designed cars utilise short remote linkages of this type built as an integral part of the gearbox or tailshaft housing (see **FIG 4:1**).

Two rod linkages with or without a housing are becoming common as a means of obtaining a more positive gearshift feel. Again these should be checked for wear at the joints.

3 Synchromesh cones and baulk rings. With the gearbox stripped down (see relevant Autobook Workshop Manual) the most important point to inspect is the condition of the synchromesh cones and the baulk ring mechanisms which are fitted on later boxes. Look for burrs on the baulk-ring dogs and ridges of wear on the cones. Also check the condition of the dogs or splines in the drive engagement mechanism. Replace parts where necessary. As this is a full gearbox stripping operation it is wise to carry this out in conjunction with the fitting of close ratio gears (see **Section 4:3**).

4 Gearbox shafts. Check all shafts for wear and examine splines carefully for signs of burring or other surface irregularites. Don't forget the splines on the input shaft and the tailshaft.

5 Bearings and bushes. Shaft alignment, noisiness and ease of change inside the gearbox are affected by the condition of the bearings or bushes in which the shafts run. Check the play of the shaft within the bearing by

attempting to move the shaft; there should be no perceptible play. Worn ball or roller bearings may permit too much end float of the shaft and although it is rare for there to be end float without side to side play this movement should be checked as well.

6 Selector rods and rails. Moving rods should slide easily in the locating holes. If they are stiff it is advisable to lap them in with fine valve grinding paste used very sparingly, followed by a final smoothing with metal polish.

The same process can be carried out on mechanisms in which the selector forks or carriages are moved along fixed rails. Always check that the wear on the selector forks is within reasonable bounds; some wear may actually aid the gear change but too much will prevent clean engagement.

7 Refitting the gearbox. It is particularly important to ensure that the gearbox alignment is not upset by the presence of dirt on the interface between box and bellhousing or at the bellhousing flange and engine joint. Gasket damage and gearbox or tailshaft mountings that are in a poor condition, or badly positioned, are frequent causes of misalignment.

4:3 Modifying the gearbox

Using close ratio gears:

In the introduction the relationship between the gear ratios, engine power and usage was discussed. In practice the manufacturer provides gear ratios for a normal small saloon that are typically:

1st	3.6:1	3rd	1.5:1
2nd	2.2:1	Top	1:1

This provides a low gear ratio for starting and, because most engines are capable of providing a reasonable pull across a wide rev range, well spaced steps of gear ratios up to direct drive in top. This is a wide ratio gear set.

There are two connected reasons why this wide spacing of the ratios is not the best for performance or competitions use. The first is that a performance tuned engine will produce significant power over a very small range of crankshaft speeds. So for maximum utilisation of this peak power engine speed, the gearing has to be arranged to give the best spread of ratios at the upper end of the speed range. Even in cars that do not have such a 'peaky' power output around a narrow speed range it is more useful to have gears that produce the highest possible speeds and acceleration around the rev/min at which maximum power is developed. The answer is to use gears with close ratios. Typical gearing would be:

1st	3.3:1	3rd	1.4:1
2nd	2:1	Top	1:1

This type of box would make a great deal of sense on a road going car, as first gear is still low enough for the car to get away on the available engine power while the other gear ratios are useful at much higher road speeds. For racing cars the ratios may draw a little closer to the 1:1 top gear.

Close gear ratios suitable for road use are available from the competitions departments of many of the major manufacturers. They are, however, quite expensive: costs range from £60 to well over £100 at current prices. Gear sets for racing can cost many hundreds of pounds.

A much cheaper source of close ratio gears for many cars is to use a box or just the gears from another car in the same model range.

Generally, high performance models in any given range have closer ratio gears. Some pitfalls to be avoided in swopping gearboxes or mechanisms are detailed in the following paragraphs.

Using a complete gearbox:

1 Check that the following dimensions, where applicable, are the same as the standard gearbox: bellhousing depth, tailshaft length, size type and position of all mountings, input shaft diameter, tailshaft oil seal diameter.
2 Ensure the number of splines on the input shaft and output shaft (tailshaft) are the same.
3 Ensure speedometer gearing is the same.
4 The gearlever should be in the same position or it may be possible to transfer the remote control linkages from one box to the other.
5 Check that the clutch release mechanism is similar or transferable.

Using gear clusters:

1 Identify the close ratio gears correctly from workshop manual and other vehicle data. Manufacturers sometimes change the gear ratios in the middle of a model run so the age of the mechanism to be used may be important.
2 Never mix gears from one cluster with those from another and avoid gear mechanisms from badly crash damaged vehicles.

Suggested close ratio conversions for British cars:

BLMC:

Mini Cooper 998 and 'S' gears will fit into the boxes on other cars in the Mini range, however, there are three types of gearbox that have to be identified carefully. Pre-1963 boxes had simple synchromesh, 1963 to 1966 boxes had one form of baulk-ring mechanism and after 1966 a better baulk-ring design was used.

The complete gearbox from the 1275 cc MG Midget (or Sprite) can easily be transplanted to the other rear wheel drive cars with the 'A' series engine like the earlier 1098 cc Midget, the A40 and the Morris Minor.

It is not so easy to find swops in the Triumph range. SAH Conversions Ltd are the specialists who can provide the best advice about gear ratios for most models in the range.

Chrysler:

Standard Hillman Hunters can take the gearbox from Sunbeam Rapier, Alpine or the Hunter GT models – the gear clusters can be swopped too. In the Avenger range it is possible to use the gearbox or the gear cluster from the Avenger GT. Hillman Imp tuners should consult specialists for closer ratio sets.

Ford:

Speedometer gear changes in manufacture are the only problem in swopping the closer ratio gears of the 1300 GT on to standard 1100 and 1300 versions of the Escort. Capri GT models have close ratios that will suit other models in the range although a different gearbox is used for the 3 litre versions which cannot be used for any other model.

Considerable problems can arise in swopping between Cortinas, Corsairs and the larger engined Escorts. Firstly, only a few of the 2000E and 2000GT Corsairs and Cortina GT's were fitted with close ratio gear clusters. These can be identified by opening the box and checking that the mainshaft third gear is larger that its adjacent selector dog. Remote control mechanisms may have to be swopped and speedometer gearing may have to be changed. Other potential problems include variation in the length of shafts and the input shaft diameter.

Vauxhall:

Viva 1256 cc models have no equivalent gearbox with close ratio gears. Some of the first Vauxhall Ventoras had close ratio sets that can be fitted to most FD series (1968-1972 Ventoras and Victors). Consult specialists for special close ratio sets.

Selector mechanisms:

Even if the parts of a selector mechanism are running smoothly it may be possible to make further gains in ease of operation by using emery paper to radius off the lips on all holes in which rods or carriages slide. For a complete reduction of friction in the mechanism special PTFE bearing bushes can be purchased in some cases for insertion into machined out holes.

Polish all sharp edges around the location points for the interlock mechanism – but not too much or the car will jump out of gear. Remove any burrs from and radius any sharp edges on the gearlever fingers that locates on the selectors.

Remote control linkages:

Older cars, like the Mini 850, had long gearlevers that were subsequently redesigned with remote control linkages. This brought the lever nearer to the driver and made it much easier to use. Various conversion kits to modify older designs of gear change to a modern specification are available. It may be possible in some instances to fit the remote change from a later model in the range. The result is not always a better gear change as the play in the linkage can exaggerate any faults in the original change. However, the gain in convenience will often make up for this.

4:4 Clutch maintenance and overhaul

Even if the engine has not been uprated to the stage where some form of clutch modification is advisable clutch maintenance is vital to the efficient operation of the gearbox.

Cable clutch operation:

Examine the run of the clutch cable and consider whether its location gives the best possible free running conditions. It may be that by routing the cable in a certain way a great deal of drag on the pedal (and cable wear) can be avoided. Graphite grease can be used on the cable.

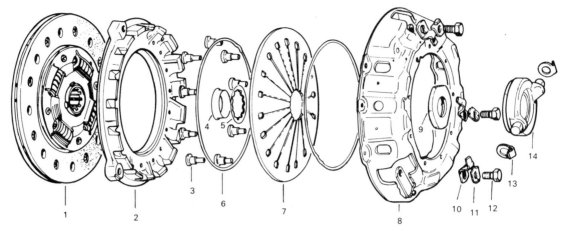

FIG 4:3 Exploded view of a diaphragm spring clutch

Key to Fig 4:3   1 Driven plate   2 Pressure plate   3 Rivet   4 Centre sleeve   5 Belleville washer   6 Fulcrum ring   7 Diaphragm spring   8 Cover pressing   9 Release plate   10 Retainer   11 Tab washer   12 Setscrew   13 Retainer   14 Release bearing

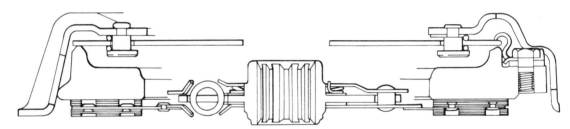

FIG 4:4   Cross-section of a diaphragm spring clutch assembly

When remounting the cable ensure that the fixed position at the bellhousing flange, to which the cable sheath is secured, is at the correct angle. It can be bent quite easily during engine or gearbox manhandling.

Hydraulic clutch operation:

Regularly inspect the condition of the clutch hydraulic mechanism ensuring that all pivots and linkages are well lubricated. Joints between metal and flexible hydraulic piping should be closely examined for leakage and the pipes themselves should be checked for splitting or abrasion damage. If there are any signs of such deterioration the components should be renewed. Changing the hydraulic fluid when the brake fluid is changed will help to prevent corrosion in the cylinders.

Clutch plate and its operation:

The clutch plate should be examined for wear and condition. If the useful friction material is worn more than half way through, the plate should certainly be replaced if it is to be used for competitions and especially if the engine power has been uprated.

The plate should slide freely on the gearbox input shaft splines. Lap in with fine grinding paste followed by metal polish if necessary.

Thrust or release bearings:

The thrust race should be in good condition, free from side to side play and end float. Check that it spins freely with no grating noises. Carbon thrust rings tend to wear unevenly and may be scored or chipped. Replace if any of these signs are found or if the carbon material has worn down to $\frac{1}{4}$ inch. The carbon material should not move in the carrier. Check that the mounting of the bearing in the clutch fork is secure and that the fork itself is free from distortion.

Refitting the clutch:

Follow clutch adjustment instructions in the workshop manual to the letter paying particular regard to any limits placed on the clutch throw out.

4:5 Modifying the clutch

It has already been stressed that the clutch must be in good condition for performance use and that for easy gear changes it is wise to overhaul the clutch operating mechanism ensuring that the release bearing is serviceable and that the clutch plate slides freely on the gearbox input shaft splines. The standard clutch maintained in this way will put up with a great deal of abuse and still be suitable for mild performance tuned engines.

However, above a certain level of engine power development considerable overloads can be placed on the clutch. In fact standard production clutches have a large leeway and will accept increases in horsepower or torque output in the order of 10 to 15 per cent with no problem other than a slightly increased rate of wear, according to use. If performance increases greater than this are made then it will almost certainly be necessary to consider one of three ways to uprate the clutch.

Older cars were fitted with coil spring clutches which have the disadvantage that at high transmission revolutions the clamping loads decrease due to the effect of centrifugal force. More modern diaphragm spring clutches act in an exactly opposite manner. The higher the speed of rotation the greater the compression exerted by the diaphragm. Thus the first possible clutch modification is to replace a coil spring clutch with a later type of diaphragm unit. This is possible with Ford Anglias and early Cortinas and it is a useful modification to the 1098 cc MG Midget and Austin Healey Sprite.

It may also be possible to add additional springs to some types of coil spring clutches. This modification should be checked with the car manufacturer's competitions department or the competitions advice department of the clutch manufacturers.

Should the power increase obtained from the engine be larger than a 10 to 15 per cent margin and hard clutch use is envisaged the next stage in uprating the clutch is to purchase a competition type from the car or clutch manufacturers. A typical uprated clutch unit will have a lining of a type with good anti-fade properties, stronger pressure plate, stronger riveting, a stiffer diaphragm spring and tougher torsional damping springs (see FIG 4:5).

Choice of the lining material for a highly tuned car is very important, although since the linings used for clutches on ordinary mass production cars have been greatly improved over the years, it may be possible to get away with using standard material of this type. One type of standard material frequently used in mildly uprated clutches is a mixture of asbestos and zinc wire. However, in severe conditions like a race start the lining temperature can rise high enough to melt the zinc. An even woven asbestos material with brass wire to conduct the heat away is the solution.

The clutches and linings described above are improved versions of the standard item. Racing clutches work on the same principles but they are lighter and stronger and do not have lining materials as such. The friction characteristics required for clutch operation are obtained by using sintered metal clutch plates with a high heat resistance. The clutches are usually designed so they can be stacked up in one, two, three or four plate units suitable for anything from small club racers to high-powered Can-Am cars.

Clutch manufacturers like Automotive Products Racing Division will advise the tuner of the suitability of a particular type of clutch for a particular application, or will supply a suitable unit on receipt of the following details: engine power output (horsepower and torque figures), maximum rev/min, diameter and number of teeth on gearbox input shaft, weight of car and type of usage (race, rally, autocross). It should be noted that uprated clutches often cost very little more than standard units. Racing clutches are, on the other hand, very expensive. The penalty of fitting an uprated clutch is a much higher pedal pressure which makes ordinary driving, particularly in towns, very tiring. This is an unavoidable disadvantage to be borne for the sake of performance.

4:6 The differential

Final drive ratios:

In addition to the variation in gearbox ratios that can be achieved further changes to the transmission can be made by the selection of a higher or lower differential ratio.

Designers of production cars apply the same criteria to the selection of the differential ratio that are used in the determination of the best intermediate ratios in the gearbox. A lower axle ratio permits a lower powered engine to cope with a wider variety of loads. More powerful engines have greater torque at low speeds and a higher ratio can be used. The car's top speed will be higher and intermediate speeds will be higher than the lower geared car at equivalent engine revs, thus fuel consumption will be kept down.

Manufacturers use a variety of differential ratios for various models within a range. Estates and vans will often have lower ratios, typically 4.4:1 or 4.2:1. GT or larger engined versions may have differential ratios in the order of 3.7:1. The tuner can play these variations to advantage.

For touring economy a higher differential ratio could be fitted provided that it is sensibly chosen. In practice this means that the ratio should not be so high as to reduce the car's flexibility in top gear. If the engine has to strain to carry out manoeuvres previously accomplished with ease, or a change down of gear is required, the savings to be made will be outweighed by the added fuel consumption at low engine revs or in lower gears. Thus the type of

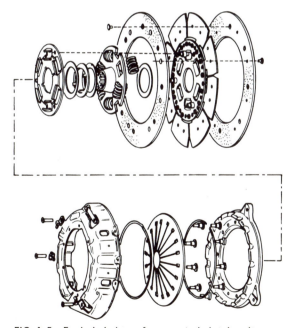

FIG 4:5 Exploded view of an uprated clutch unit

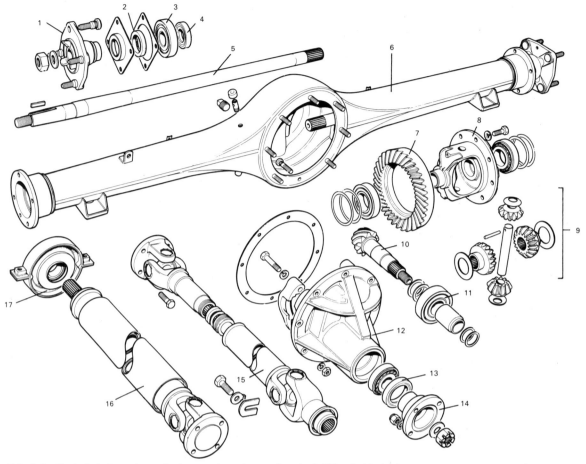

FIG 4:6   Exploded view of a typical rear axle and propeller shaft (Morris Marina)

Key to Fig 4:6   1 Hub   2 Seal   3 Bearing   4 Seal   5 Halfshaft   6 Axle casing   7 Crownwheel   8 Differential cage   9 Differential pinion (star wheel) assembly   10 Pinion and shaft   11 Pinion bearing   12 Gear carrier or nose piece   13 Seal   14 Drive flange   15 Propeller shaft, rear section   16 Propeller shaft, front section   17 Propeller shaft centre bearing

usage has to be borne in mind when deciding on the ratio to be used. Long distance commuting on motorways or dual carriageways where high constant speeds can be maintained might be accomplished more cheaply by using a higher gear ratio for many small cars. The benefits diminish if more motoring is carried out in traffic or on roads which will not allow constant high speeds to be attained.

Small ratio changes can make quite a large difference in economy provided a suitable differential unit can be obtained cheaply. The only way of doing this is to look at the specifications of models with a similar rear axle or differential housing and finding one with an appropriate ratio. For example, the Mk 1 Ford Escort 1100 is produced with a 3.9:1 differential ratio and it is possible to fit the 3.777:1 ratio unit from the 1300 cc version. This can be obtained from a scrapyard quite cheaply.

The car's performance characteristics change more than its fuel economy with a swop of differential ratio. A low ratio will result in a lot of revs and a low top speed but it will also transform acceleration times. Like the use of

close ratio gears, it may also permit much effective use to be made of a performance engine's maximum power peak. Using the Escort example above, a possible swop is to use the 4.125:1 ratio from the 1300 cc van and estate to replace the standard 3.777:1 ratio of the saloon.

It should always be borne in mind that a change to the differential ratio affects the overall gear ratio in all gears. For example, in the wide ratio gearbox mentioned above coupled to a 4.2:1 differential, the overall gear ratios are as follows:

	Intermediate gearbox ratios	Differential ratio	Overall ratios
1st	3.6:1	4.2:1	15.12:1
2nd	2.2:1	4.2:1	9.24:1
3rd	1.5:1	4.2:1	6.3:1
Top	1:1	4.2:1	4.2:1

But fitted with a 3.7:1 differential the overall ratios are:

1st	13.32:1	3rd	5.55:1
2nd	8.14:1	Top	3.7:1

It is a hypothetical combination that could have several advantages, like the extended speed range of first gear (note this is similar to the use of a close ratio bottom gear) and, provided the gearing is not too high, lower engine revs at high road speeds. The drawback will be the loss of acceleration in 2nd and 3rd gears. If the car were designed for a 3.7:1 differential and a 4.2:1 unit were fitted there would be better 2nd and 3rd gear acceleration perhaps to the point where no useful speed could be gained before the engine reached maximum power. The rule is to make the smallest ratio change possible always bearing in mind the ratios which the manufacturer has used for each of the various models in a range.

Practical aspects of differential swops:

In many model ranges it is possible to fit at least one size larger or smaller differential into the same housing. However, on cars for which no alternative differential is provided the only answer may be to swop the entire axle unit. In some instances this is a relatively simple bolt-on replacement job. On other cars shortening or lengthening of the prop shaft may become necessary, combined with modification to the means of attachment at the suspension. Where a complete axle is used, one of the most important points to check is that any change in wheel spacing will not affect clearance of the wheels within the wheel arches.

Further differential points to watch:

1 Examine tooth wear on crown wheel, pinion and star clusters and abandon if any chipping or burring is found.
2 Renew all gaskets and oil seals.
3 Observe the tighening torques detailed in the workshop manual.
4 If a new differential unit is being fitted fill the casing with a special grade of hypoid gear oil used for initial lapping in of the gear teeth. Older units should be filled with the recommended grade of hypoid lubricant.

Limited slip differentials:

The normal differential is a means of allowing the driving wheels to revolve at different speeds while describing circles of differing radii during cornering. It is an excellent and generally very reliable device but it displays one serious disadvantage in certain conditions. If one wheel loses traction on grass, mud, ice or in hard cornering under power, the differential behaves in the same manner as on a corner and allows the free wheel to race while the other stands still.

The answer is a limited slip differential which utilises a system of clutches in such a way that however free one wheel is to slip there is always some drive transmitted to the other wheel. The device is an expensive addition to the car's transmission equipment and should really only be considered as a competition modification.

4:7 Other transmission components

Universal joints:

A number of universal joints are used in the transmission system to allow vertical movement of the sprung transmission components. These items are placed under great stress by certain types of usage, e.g. racing starts. Like other parts of the system they should be maintained in as good condition as possible. Check each joint by attempting to lever each yoke in turn away from the spider. Any perceptible movement will mean an overhaul of the joint is required. A few modern joints are machine assembled and cannot be repaired – these should be replaced by a whole new propshaft. If the possibility of rapid joint wear is high, manufacturers like Hardy Spicer can be consulted to provide a suitable replacement for joints of this type.

Some joints like the BLMC pot joints on a Maxi are covered by a rubber gaiter packed with a special grease. Make regular checks to ensure that the rubber is in good condition and that there is no leakage of the grease or ingress of dirt. Dirt and grit will cause very rapid wear in such joints.

Propeller shafts and half shafts:

Three problems arise with prop shafts. Firstly, during the design of a performance car it may become necessary to construct a shaft of a specific non-standard length to suit a particular combination of engine, gearbox or rear

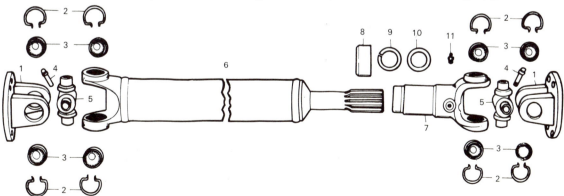

FIG 4:7   Components of a propeller shaft assembly

Key to Fig 4:7   1 Flange yoke   2 Circlips   3 Cups   4 Grease nipple   5 Spider   6 Propeller shaft   7 Sliding yoke
8 Dust cap   9 Steel washer   10 Washer   11 Grease nipple

axle. The cutting and welding work should be carried out by a machine shop with adequate balancing capabilities – again Hardy Spicer will be able to help. The balancing of a standard shaft may become critical in some applications and expert advice should always be sought to overcome this problem. The final point about prop shafts is that the splined sliding joint which allows for lengthening of the gearbox/rear axle distance in suspension travel should be regularly inspected for spline condition and freedom to move.

Half shafts are particularly vulnerable components in rallying, racing and autocross in which savage changes in load are applied with great frequency. Several car manufacturer's competitions departments offer shafts made from stronger materials which are, like torsion bars, crack tested.

CHAPTER 5

Tuning the suspension

5 : 1 The function of the suspension
5 : 2 Describing handling characteristics
5 : 3 The aims in suspension modification
5 : 4 Shock absorbers and dampers

5 : 5 Renewing, replacing and adding dampers
5 : 6 Lowering and stiffening the suspension
5 : 7 Other suspension aids

5 : 1 The function of the suspension

Cars are one of the ultimate luxuries of the consumer society and one of their most alluring characteristics is the comfort with which long journeys can be undertaken. The term comfort, related to the car, depends on a number of factors including sound insulation, the nature of the seating and the efficiency of the heating and ventilation system. But by far the greatest influence on comfort is exerted by the car's system of suspension.

The suspension of a standard car is designed to soak up the rough and tumble of the road surface, reducing the shocks imparted to the wheels by potholes, bumps and stones and insulating the passenger compartment from excessive undulation on surfaces of varying pitch and level. The suspension system has far reaching affects on the car's steering, roadholding and handling characteristics. In addition the rear suspension of conventionally laid out cars (front engine, rear wheel drive) has the task of locating the rear axle, coping with the weight of the transmission components and bearing the forces of acceleration under power.

These are just a few of the factors which make suspension design one of the most complex automobile engineering subjects. It suffices to say that on a car in which the performance has been uprated from standard, parallel improvements will almost certainly have to be made to the suspension characteristics to enable the most efficient and safe use of the power increase. Improvements to the car's standard suspension can bring superior handling during cornering, reduction of violent transitional characteristics resulting from the effect of cornering forces on the wheels and car body, better transmission of the engine's power to the road and, to some extent, enhanced braking characteristics, all conducive to safer and more pleasant motoring.

Caution in suspension modification:

As with all tuning preparation on cars you cannot alter one suspension characteristic without detracting from another. Three aspects of the suspension can change drastically as alteration of the standard car's suspension is undertaken. First and foremost is that the comfort factor carefully built into the car by the manufacturer will almost certainly be changed for the worse.

Human judgement of comfort in cars, related to the suspension, is very subjective. What is one man's meat is another's poison. Many men, all too often the only driver in a family, enjoy a feel of the road through the seat of their pants. This is often variously described as a 'firm ride', 'taut feel' and so on. A certain degree of feedback from the wheels gives the driver an impression of what the car is doing and reassures him that he is in control. However unskilled the driver may be, there is a sense of insecurity with an over-insulated ride simply because there is a lack of sensory stimulation.

Passengers on the other hand wish to be well insulated from road shocks but without the floating feeling of isolation from the environment that can lead to car sickness.

Not all cars achieve this balance between insulation and 'feel' of the road very successfully. Many larger models from European manufacturers and most American cars are excellent for motorway cruising. However, on the chop and change pitch, camber and corner patterns of smaller roads the characteristics of the suspension which endow a boulevard ride become a distinct disadvantage. The car overreacts to every condition. Roll becomes exaggerated and undulations introduce an uncomfortable floating sensation, particularly at the rear of the car.

Mid-range European cars almost all have the family mix of characteristics favoured by both driver and passenger. Smaller saloons, as a by-product of their lower mass, shorter length, cheapness and lower maximum speeds usually give a much harsher ride.

The other factors that often suffer in suspension modification are the noise insulation and, what is to some extent a related feature, transmission of vibration and jarring from the road into the passenger compartment.

These debits to suspension modification add up to a caution – the altered characteristics cannot be expected to please everyone in the family. Like the balance that must be observed in making improvements to the engine's power it must always be borne in mind what purpose the car is really intended for.

The corollary to this caution is that there is absolutely no point in making suspension modification for economy ends. Work of the kind discussed in this chapter is wholly concerned with allowing the car to go faster with a greater degree of control and ultimately, safety. That is not to say that some mild suspension tuning cannot make a great deal of difference to the occasional pleasures of putting a foot down and ignoring economy measures.

5:2 Describing handling characteristics

The road tests of cars published in the motoring magazines all refer with varying emphasis to the car's handling characteristics and give subjective impressions of the roadholding ability. The ordinary motorist could be forgiven for understanding very little of the jargon used. Things like understeer, oversteer, throttle steering, tail-out cornering and axle tramp seem to have little to do with everyday motoring. But these terms do describe important aspects of a car's behaviour and some acquaintance with them is essential to an understanding of the value of possible modifications. The following are descriptions of some of the major factors involved in handling and roadholding, all of which become more important as the performance is increased.

Understeer and oversteer:

When a car is steered round a corner, considerable forces are exerted on the tyres, particularly in the area where the tyre tread contacts the road. The wheels are pointed along the intended curved path, but the mass of the car wants to go straight on (it has inertia). The resultant conflict of forces distorts the tyre, with the result that the tread contact patch on the road follows an actual path a little nearer the 'straight on' line than the steered course. The angle between the intended and actual tyre paths is called the slip angle (see **FIG 5:1**).

The term is not intended to imply any loss of adhesion between tyre and road; tyres are designed to provide a great deal of grip in these conditions. What occurs is a kind of squirming or creeping of the tread on the road, combined with distortion of the tyre sidewalls. Skidding or sliding, which results from loss of adhesion, is a different thing altogether.

Slip angles begin to occur as soon as a tyre is deflected from a straight line, although they are very small at low speeds. Once the car is following a curved path, both front and rear tyres generate slip angles, and the balance between front and rear slip angles determines whether the car understeers or oversteers.

If the front slip angles are greater than those at the rear, the car understeers. In practical terms, the driver has to turn the steering wheel more and more into the corner (more lock) to follow a constant radius turn, the faster the car is travelling (see **FIG 5:2**).

The opposite condition is oversteer, when the rear slip angles are greater than the front so that the tail of the car tends to run wide on the corner. The practical response to this tendency is to back off the steering wheel (less lock) to keep the car on a constant radius (see **FIG 5:3**).

Neutral steering, midway between understeer and oversteer, rarely occurs over any very wide range of cornering speeds. But it may be encountered during the gradual transition from one extreme to the other which takes place with some cars at certain speeds on some corners.

So understeer and oversteer are products of the balance between front and rear slip angles, and the slip angles are products of tyre distortion. It follows that the car's behaviour in these terms will be determined by the factors affecting tyre distortion, and the most obvious of these is

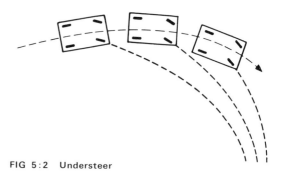

FIG 5:2 Understeer

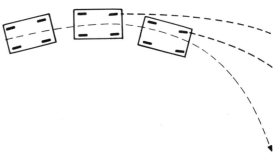

FIG 5:3 Oversteer

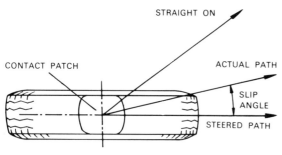

FIG 5:1 Slip angle

the car's weight distribution. As a general rule, front heavy cars – the majority of cars, since they are front engined – will tend to understeer, while back heavy cars – rear engined cars, loaded estates – will tend to oversteer.

But there are exceptions to this rule, because there are many other factors affecting tyre distortion, factors as simple as tyre pressures (a softer tyre will run at larger slip angles than a harder one) or as complicated as the effective geometry of the suspension layout on the car in question. In practical terms the most important points are that the car should not exhibit an excess of either understeer or oversteer and that its behaviour should be reasonably consistent and predictable.

Throttle steering:

One of the factors affecting the balance of a car during cornering is the amount of power being delivered through the driving wheels. Consequently, any change in the throttle opening can influence the handling both because of the resultant weight transfer and its effect on suspension geometry and because the amount of torque transmitted through a tyre conditions the amount of grip available for cornering. This effect is known as throttle steering – with the steering wheel held in a fixed cornering position the radius the car describes around the corner can be varied with the throttle pedal – and it is much more marked on some cars than others. Some front wheel drive cars, for example, have a noticeable 'lift off, tuck in' characteristic when cornering hard – easing the throttle tightens the car's line round the corner, a useful attribute providing it is not too abrupt.

Roll and the roll centres:

All cars have a tendency to roll on corners, some more than others. Rolling is not disadvantageous to performance provided that the movement of the sprung mass of the body does not lift any of the wheels off the ground

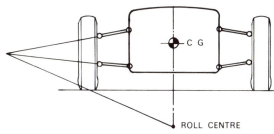

FIG 5:4   Roll centre and CG: wishbone suspension

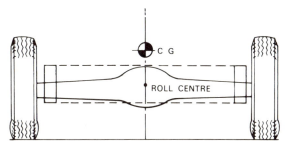

FIG 5:5   Roll centre and CG: live axle suspension

resulting in loss of traction or a violent change in steering characteristics. Whether or not the roll upsets the delicate steering and traction balance on a corner depends largely on the location of the vehicle's roll centres.

The roll centres are the points in the planes of the front and rear suspension about which the sprung weight of the car (this includes the driver, engine, body and some of the heavier components of the transmission) rotates under the influence of cornering forces.

The suspension exerts a resistance on the sprung weight of the car in its tendency to roll which is referred to as the roll stiffness. When the front suspension stiffness varies from the rear suspension stiffness, for example, in a car with independent coil spring front suspension and a rigid rear axle on leaf springs, more of the resistance to roll is accepted by the stiffer assembly. Consequently, the effective transfer of weight from the inner to the outer wheel during cornering is also greater at the stiffer end of the car. The result in the example cited is that the inner rear wheel will lift off at extreme cornering forces, causing wheelspin and loss of traction. One car which displays this characteristic, quite safely and gracefully, is the Chrysler Avenger a car on which conventional suspension design is taken to very advanced limits of safety.

But the extent to which any car is able to roll is also determined by the relationship between the roll centres and the centre of gravity. The higher roll centres of non-independently sprung vehicles, resulting in the pronounced roll stiffness effects described above, can be balanced to some extent by lowering the car's centre of gravity. Cars with all-round independent suspension generally have lower roll centres, but lowering the centre of gravity is still advantageous in the quest for greater cornering power.

Axle tramp:

One of the penalties of increasing the power of an engine is that various other systems of the car are not able to cope with the extra loads or the higher speeds. This is certainly true at the point where the power reaches the road – for many cars the critical area is the rear axle and the rear suspension particularly when leaf springs are employed.

In accelerating under maximum power from rest or from low speeds the engine, transmission and rear wheels are fighting against the vehicle's inertia and the rolling resistance. It is possible for many cars, modified or not, to lay down enough power to defeat the traction system under certain conditions but the tendency is increased on performance cars. The problem is that the engine can provide more power than the car's rear end can cope with in attempting to propel the car forward. Some energy is lost in wheel slip but a lot more is diverted into winding up the springs. Wheelspin suddenly provides a chance for the energy stored in the springs to escape and the result is a violent kick of the axle. The cycle is repeated with some rapidity and violence causing an effect known as axle tramp.

Axle tramp results in a loss of power transmitted to the road but more importantly it betrays a fundamental weakness in the rear suspension. Axle patter is the name given to a similar effect when it happens on corners – with a consequent dangerous loss of traction and grip.

Tramp and patter can also result in damage to suspension components, axle casing and half shafts. The cause of the trouble is inadequate location of the axle, so stiffening the springs or, better, providing additional location bars will overcome it.

Wheel tramp and patter can also occur during hard acceleration on some front wheel drive cars. But rear wheel drive cars with independent rear suspension are usually immune.

5:3 The aims in suspension modification

The major objectives in modifying or tuning the suspension are to improve the roadholding, the car's ability to sustain cornering forces and retain its adhesion to the road surface, and to endow the handling with greater predictability and controlability – to quell any tendency for an abrupt transition from understeer to oversteer, for example – so that the roadholding can safely be exploited. To achieve these aims measures like lowering the ride height, increasing the roll stiffness of the front or rear suspension and improving the effectiveness of the dampers are employed. Fitting wider wheels and tyres, discussed in the next chapter, can play a part too. What is required will vary greatly from one car to another depending on its original characteristics and the degree of increased performance obtained by other modifications.

Caution:

No suspension modification should be carried out unless the remaining standard components in the suspension are in good condition. Particular attention should be paid to the following points:

1 Suspension mountings. The metalwork of the body or chassis around all the suspension mountings must be free from corrosion which would affect its strength. The localised loads can be very high and modifications may increase them. Mounting rubbers or bushes should be in good condition; it is always wise to consider replacing them with uprated parts (harder rubber or steel inserts are used) if they are available.

2 Ball joints, kingpins, wheel bearings. Check these for excessive play. In practice there should be no perceptible wear in these components when the wheel assembly is shaken by hand with the car jacked-up.

3 Leaf springs. Clean leaf spring units as well as possible and check for cracked leaves.

4 Shock absorbers or dampers. If the standard units are to be retained as part of the tuned set-up check that they are in good order.

5:4 Shock absorbers and dampers

Although the major characteristics of suspension systems have been described nothing has been said so far about how these effects are achieved. Modern suspensions are a partnership between the suspension linkage, the springs and the hydraulic devices that are often referred to as shock absorbers – more correctly they are dampers.

It is possible to make tuning modifications to all parts of the car's suspension but for immediate gains at a reasonable price the enthusiast very often attends to the dampers first.

Various types of damper are in common use on cars and they all utilise a common principle. Secured between the car body and the unsprung components of the vehicle (usually attached to the axle or wheel hub assembly) they damp the oscillation of the springs when the bump absorption action of the suspension has taken place. The simple theoretical damper offers little resistance to the bump or compression movement of the spring and a great deal of resistance to the spring's rebound action.

This means that the spring is ideally only allowed to oscillate through one rebound from full compression and a half stroke down to normal ride height again. But as far as the design of the damper and spring system is concerned the reality may be far from the simple theoretical pattern described above. For instance it may be necessary under some circumstances to provide some resistance to the bump stroke of the spring or to vary the ratio between the bump and rebound damping.

There are several main types of damper in use on modern cars. The two most common are the telescopic type and the unit which forms the central core of a MacPherson strut assembly. These are similar constructed units filled with hydraulic fluid (sometimes filled under gas pressure). Under compression the fluid flows from one chamber of the unit to the other and valves or precision orifices offer resistance to the return flow during the rebound stroke. FIG 5:6 shows a telescopic damper in cross-section.

Most original equipment telescopic units are non-adjustable; accessory types may be 25 to 30 per cent uprated or adjustable. The element of adjustment varies from rebound only (Koni) to fixed ratio bump rebound (Armstrong and Spax). Adjustment may be by means of

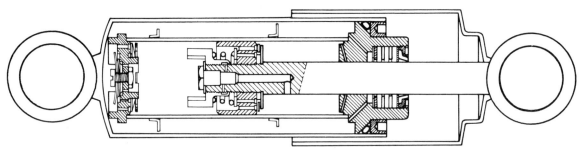

FIG 5:6 Cross-section of a Koni telescopic hydraulic damper

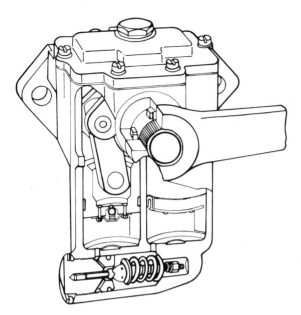

FIG 5:7 Cutaway view of a lever arm damper

an external screw or by undoing one end of the unit and compressing the two halves together to engage an internal adjuster; rotating one half of the unit about the other then changes the rating.

On many older cars the rear dampers are lever units in which the radial movement of an arm activates a piston in a hydraulic cylinder. The movement of the hydraulic oil is restricted or freely passed by valves or orifices in the same way as in telescopic types. Similar in action to the lever type is the piston shock absorber which is designed as the upper wishbone of an independent front suspension assembly. Uprated lever arm dampers are available but often the best way to uprate suspensions with this type of damper is to replace them with telescopic units. Piston dampers which form an integral part of the suspension system can be supplemented by telescopic types or made inactive by removing the valves.

Motorists are only just beginning to realise that dampers like many other components of the car, are to be considered as expendable items. Working in the worst conditions, under the wings, exposed to dirt, salt and water (and working very hard) they can quickly become ineffective. The result is a lowering of the level of safety originally inherent in the car's handling. Controlability and ride suffer and the car can become very unstable under heavy braking. The problem is that the deterioration is gradual and hence tends not to be noticed by the owner driver. Generally dampers should be replaced at intervals of no more than 30,000 miles.

Hard and fast driving over rough surfaces greatly increases the work that the dampers have to perform and thus shortens their life – it is not uncommon to find cars with well under 10,000 miles on the clock requiring new dampers. Causes of damper deterioration include fluid loss, fluid contamination, wear and corrosion, valve failure and, during operation at speed, aeration of the hydraulic fluid.

The simplest test of damper operation is to bounce the car up and down by pressing on the wings, releasing at the bottom of a downstroke and checking that the car comes to rest after one upstroke and half a downstroke. The test should be repeated pressing each wing down in turn.

This test is by no means definitive and may only show up dampers in the worst possible condition. A more exacting test can be carried out by many garages to a procedure developed by the Shock Absorber Manufacturers Association (SAMA). The SAMA test rig consists of a short ramp with a release mechanism which drops the car a few inches. A pen and recorder mechanism traces the movement of the car body after the drop. Front and rear wheel pairs are tested separately. The only drawback to this system is that it may not differentiate sufficiently between the performance of the dampers and that of other components in the suspension.

Choosing dampers:

A wide range of telescopic dampers are sold by tuning shops and mail order firms advertising in the specialist performance magazines. Armstrong, Girling, Spax, Koni and Woodhead are well known makers offering ranges designed to fit most popular cars.

The first decision to be made is whether to go for adjustable or simply uprated units. The latter are generally cheaper and when carefully selected for the front or rear of the car, or for both ends, depending on the characteristics of the vehicle in question, they can make an enormous improvement in the car's handling.

Adjustable units, particularly those that can be used to vary the suspension's bump resistance as well as the rebound characteristics, have two advantages. Foremost is the ability to tune the suspension to prevent rapid pitching and rolling movements in short, tight corners and S-bends. The dampers alone will not prevent the car rolling on a long, fast bend, but the shorter term effect can

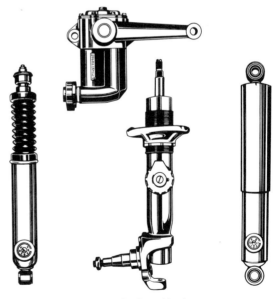

FIG 5:8 A selection of adjustable dampers

be beneficial. The second advantage is one of economy; as the damper unit wears the settings can be adjusted to compensate for the loss in performance, lengthening its effective life.

5:5 Renewing, replacing and adding dampers

Damper units have two types of mounting. It is common for the lower mounting to be a rubber bushed pivot. The top mounting can be a second pivot or a washered bolt. Upper mounting points are generally easily accessible in the boot, under the wings or on the flitch panels inside the engine compartment. Lower mountings are on the axle casing or the wishbone, kingpin or wheel hub assembly.

Ensure that replacement units with the correct type of mountings are purchased. Check that the mounting areas for the units are capable of withstanding the forces that will be applied. These forces are, of course, nowhere near the full weight of the car. Usually body panelling in good condition is sufficiently strong to accept the loadings of the stiffest units but watch out for corrosion of panelling under the wings. Always replace rubber bushes and washers carefully noting the position of flat and cupped washers in relation to one another.

In fitting additional telescopic units adequate brackets must be fabricated from mild steel. Brackets should be welded into place on the lower mounting position but bolting will do at the upper position (useful if the units are to be removed at a later date).

Particular attention must be paid to the angle of the new damper units. The action of the suspension almost always allows wheel movement through a shallow arc and the mounting angle of the telescopic unit should be at a tangent to that arc. In the case of a trailing arm the unit would be upright on the car's longitudinal axis but inclined slightly towards the front of the car. In the case of rear leaf springs where radial movement is limited the unit should be at a very slight angle inwards and need only be pivoted at the lower mounting. Adding dampers to independent suspension systems requires much more attention to the arc of the wheel's swing. Units should be

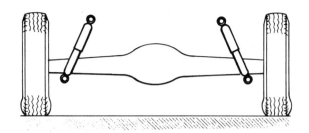

FIG 5:10 Dampers on live axle suspension

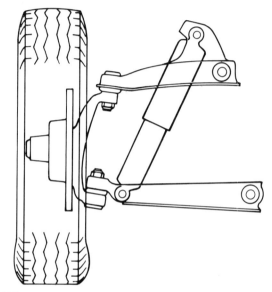

FIG 5:11 Damper on wishbone front suspension

pivoted at both ends ensuring that the pivot axis is aligned with the car's longitudinal axis.

Dampers mounted inside springs are more difficult to deal with. Often it is necessary to use a suspension spring compressor to take the spring load while removing and replacing the damper unit. This is because at full extension the damper maintains the spring in considerable tension. Good quality spring compressors, either triple hooks which loop over the spring coils or adjustable tools similar to G-cramps are available at tuning shops. Some manufacturers offer special tools to cope with particular suspension types – these can be bought or hired from a tool specialist.

Extremely high spring tensions are in use on most suspensions. Never use sub-standard spring compression equipment or rely on home made tools or methods. Spring coils clamped under tension should always be secured in at least three positions equally spaced radially around the spring.

Replacing damper units in MacPherson strut assemblies:

The best way to uprate MacPherson strut units is to replace the original struts with a complete new assembly having a lowered and stiffened spring and a stiffer

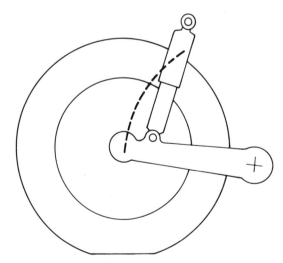

FIG 5:9 Damper on trailing arm suspension

damper unit. However in cases where better shock absorption only is required it may be possible to obtain a 25 or 30 per cent stiffened damper only. For competition purposes use gas-filled struts of the Girling or Bilstein type; this type of strut is immune to fluid aeration which can otherwise occur in extreme conditions.

Replacing the damper unit in a strut entails similar precautions to those used when tackling a damper inside a spring. Special spring clamping tools should be used and reference should be made to the workshop manual to see if strut dismantling can be achieved on or off the car. Once the strut is dismantled, uprated dampers are straight replacements for the standard units.

While the strut is off inspect the upper mounting area for signs of buckling or corrosion – Fords in particular are noted for weakening of this area and it is certainly advisable that reinforcement plates are welded in if competition use is envisaged (regulations permitting).

Certain types of MacPherson strut have rubber upper bearings which can be distorted during replacement. Workshop manuals give a proper procedure for tightening to prevent steering bias. Replace all sealing materials when refitting the strut, particularly on units with ball or taper bearings as water entry and corrosion of the bearings is a frequent problem.

5:6 Lowering and stiffening the suspension

The car's centre of gravity (therefore the influence that cornering forces can exert on roll) can be lowered in a variety of ways according to the specific type of suspension system. But before describing some of the methods used there are a few general points that must be considered:

1 Lowering the suspension without stiffening the springs (altering the spring rate) will mean that the suspension will bottom more frequently – the lower the modification the greater this problem will become. Stiffer springs overcome this effect to some extent. An alternative is to use some form of spring assister or a bump stop with progressive arrest characteristics (see **Section 5:7**).

2 Check that there is adequate clearance between the wheel and the wheel arches under full suspension compression and on full lock especially if wider wheels are used (see **Chapter 6**).

3 Check that there is enough clearance under the rear of the car and in the transmission tunnel to accommodate the rear axle, differential and prop shaft where appropriate.

4 As the bump stops are likely to be called into action more often, check that the bodywork around the stop mountings is capable of taking the extra strain.

Rear suspension modifications:

Leaf springs. The easiest way to lower leaf spring systems is to use a spacer block between the top of the spring and the axle. For a normal road going car a maximum reduction in height of $1\frac{1}{2}$ to 2 inches is permissible. There are many kits on the market to perform this operation; all are relatively simple to fit. The standard U-bolts are removed, the transmission is jacked-up to give a suitable clearance for insertion of the blocks and new longer U-bolts are secured in place of the original fastenings.

There are other lowering possibilities with leaf springs. The springs themselves can be flattened by a suspension specialist or flatter springs can be bought. It may also

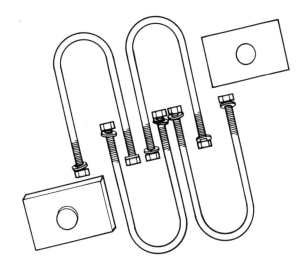

FIG 5:12   Rear axle lowering kit

be possible to invert some springs of the single leaf type. This trick should only be carried out if: **1** the springs are in excellent condition; **2** the springs are of a very rigid almost flat type; **3** the hangers and shackles have sufficient clearance to accept the reversed pivot eyes.

Stiffening or uprating leaf springs is a relatively easy modification. Uprated springs are available from tuning specialists. There are also various assisters on the market (see **Section 5:7**). Perhaps the cheapest way of finding stiffer springs is to use the generally heavier duty types produced for the van or estate versions of some cars. These may present the possibility of a simple bolt-on modification; check with a dealer or main agent if the springs will fit a saloon.

Do not attempt home modification of the leaf spring assembly. Modern steels are used to make the spring leaves and these do not respond to crude heating and

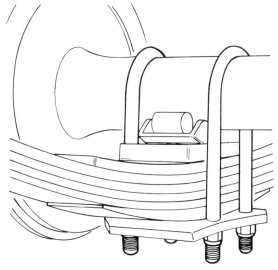

FIG 5:13   Rear axle lowering kit installed

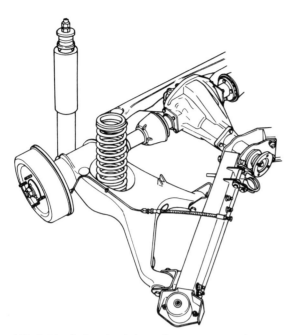

FIG 5:14 Coil spring independent rear suspension

forging processes – tempering has to be carried out by specialists who can gauge temperature treatment to within a degree or two of the desired value. Remember that handbrake cables will have to be retensioned when the suspension is in its lower position.

Coil spring trailing arm, semi-trailing arm and de Dion types. Rear suspensions incorporating a coil spring vary in design and complexity – usually they are far more difficult to work on than the leaf spring type. Theoretically lowering these types of suspension is simply a matter of shortening the springs. In practice a number of difficulties can arise from this process.

Like leaf springs, coil modification is a job for experts – it is simply not good enough to hack a section off the spring to obtain enhanced handling. This would leave a spring of poor tension characteristics. Far better is to fit shorter replacements or have the springs heated and squeezed to a shorter length. This has the added benefit of increasing spring stiffness.

Second best is to have the springs professionally shortened by removal of a portion of the coil. Before refitting a shortened spring, check that the cut end is free from burrs and that sharp edges are flattened off – otherwise the spring could eventually wear a hole through the locating point. A spring treated in this way must have some means of additional stiffening applied to it – the pressurised ball type (Auto-balans) described in **Section 5:7** is a good choice.

To remove rear suspension springs it will usually be necessary to clamp the springs under compression in the same way as described in **Section 5:5**. Other points to watch with these types of suspension are:

1 Rubber bushes. Most coil spring rear suspension designs rely on a number of rubber bushes in the suspension members and torque rod links to provide varying degrees of compliance and, very often, insulation from road noise. The bushes may be pivoted eccentrically or may have irregular cut-outs to meet these requirements and it is essential that they are mounted the right way round in the various mounting eyes. Never use engine oil or grease to make bush insertion easier as it damages the rubber. Instead use detergent and water or brake fluid. Special tools are often necessary to insert rubber bushes. Whenever possible the tools should be hired as the bushes are relatively delicate components which play an important part in the overall suspension characteristics of the car and makeshift methods may damage them.

Special tools are invariably called for in the removal and replacement of suspension arm roller or ball bearings and bushes. Plain metal bushes may have to be reamed to size in situ when replaced. Where possible avoid these complicated procedures by regular grease lubrication, inspection of protective rubber gaiters and ensuring that no grit or dirty grease enters the bearing during suspension modification.

3 Remember that the suspension may be slightly biased – the righthand spring may be longer than the lefthand spring. A colour code is often used to distinguish between the two springs.

Front suspension:

MacPherson strut. The MacPherson strut design of suspension is widely used in modern mass production cars because it is a cheap compact solution to three problems – suspension, damping and pivoting for steering.

Uprating the damper unit contained within the strut has already been discussed in the damper section. As

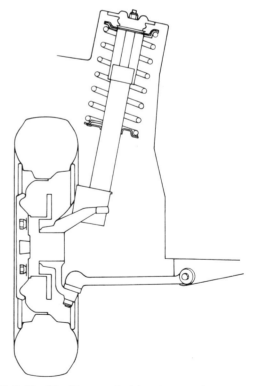

FIG 5:15 MacPherson strut front suspension

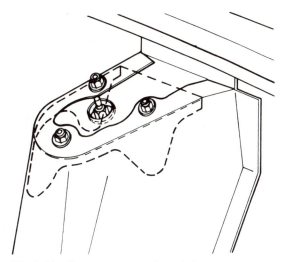

FIG 5:16   Upper strut mounting reinforcement plate

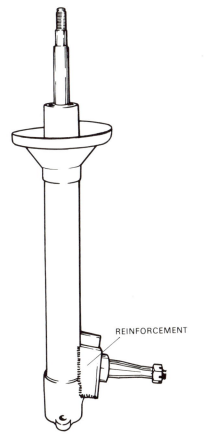

REINFORCEMENT

FIG 5:17   MacPherson strut with welded-in wedges

with all coil spring devices the lowering is accomplished by shortening the spring. For many popular small cars like the Ford Escort, tuning specialists have ready made uprated strut assemblies which can simply be bolted on

in place of the standard unit. Failing this the best solution is to have the spring professionally shortened and uprated.

Strengthening both the strut and the mounting area is an important consideration for certain types of competition use. It is normal to reinforce the area of the joint between the stub axle and the strut with welded-in wedges. Extra plates can be welded into the top of the wing where the strut meets the body. The ultimate strengthening is achieved by steel brace bars spanning the engine compartment from the top of one strut to the other. Alternatively the braces may be angled back to the centre of the engine compartment bulkhead.

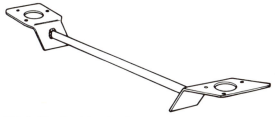

FIG 5:18   Strut bracing bar

Double wishbone or coil-spring parallelogram suspension. Used in many forms for the front suspension on small cars, the double wishbone type consists essentially of a coil spring sandwiched between the body or subframe and the lower of two swing links or wishbones. A swivel axle is ball-jointed top and bottom to the wishbones. Most examples of this type of suspension have a telescopic damper mounted to the lower wishbone through the coil spring to the bodywork. A piston type damper may be used operated directly by the upper wishbone.

Lowering the suspension is carried out by using shortened springs available from specialist tuning equipment suppliers.

Spring compression tools are required to dismantle this system and it is usually necessary to have impact wedges or a screw tool to break the taper joints on the ball joints. Special points to watch are:

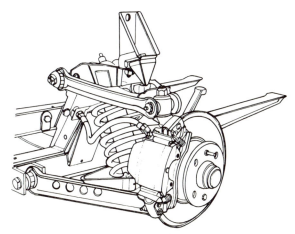

FIG 5:19   Wishbone independent front suspension

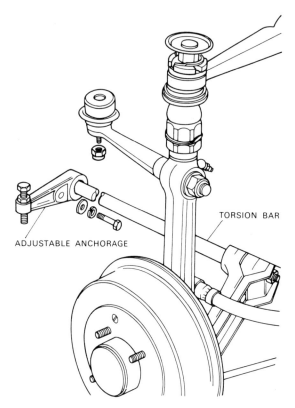

ADJUSTABLE ANCHORAGE

TORSION BAR

FIG 5:20   Torsion bar front suspension

1 The lower wishbone is often made from pressed steel and cracking around both inner and outer pivots is common. Look for corrosion of this member, especially around the spring seats, as it is exposed to a lot of salt, dirt and moisture.

2 Bottom swivel joints are particularly prone to seizure – check the condition and renew if necessary. Lubricate both joints regularly if greasers are provided.

3 Rubber bushes are used in this system for the same reasons as in the rear trailing arm types – compliance and insulation. Check that they are in good condition and replace if necessary taking care not to split the rubber during insertion. Worn rubbers can seriously affect steering geometry and cause irregular tyre wear.

4 Front suspension bottoming can become a problem with this system and it is wise to use progressive bump stops in addition to the uprated springs.

Torsion bar suspension:

Torsion bar suspension is often considered old fashioned but it does have many excellent characteristics, not the least of which is the ease with which the ride height can be adjusted. A torsion bar is a length of special steel rod. It is anchored to the chassis at one end and a suspension member is secured to the other end. The twist of the bar resists the compression forces applied to the suspension.

Various types of torsion bar adjustment are provided – the best is the Volkswagen type which utilises the difference in number between the fixed end locating splines and the suspension end splines to provide a wide variety of fine adjustment positions. Other types of adjuster are more limited and are often only intended as production tolerance adjusters or to take up the 'setting' that occurs as the bar ages.

Consult the particular car's workshop manual for specific torsion bar adjustments but bear the following points in mind:

1 Progressive adjustment by means of a bolt-on lever is provided in some systems but other types may have regular interval settings for raising or lowering the suspension – the workshop manual will say what each adjustment means in terms of body lowering.

2 Check that the torsion bar is in good condition. Nicks or deep scratches can be the starting point of failure in this highly stressed component.

3 When replacing a torsion bar set up the suspension $\frac{3}{8}$ inch to $\frac{1}{2}$ inch too high to allow for the 'setting' of the bar under load.

4 Generally with the wheels hanging free, all load is taken off the torsion bar so it is perfectly safe to dismantle in this position, but this point must be checked in the workshop manual.

5 Setting up this system is usually a trial and error procedure. Ensure that the damper is not interfering with the settling position during measurements. Disconnect the damper or bounce the car up and down several times to get a correct reading. The suspension at the other end of the car can also interfere with this procedure. Check that there are no ride height irregularities at the rear wheels with the front wheels jacked-off the ground before starting to measure at the front.

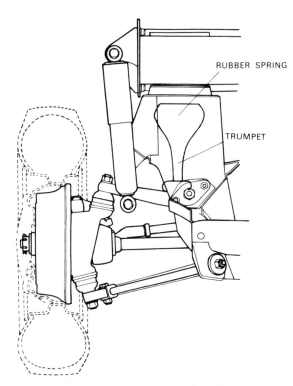

RUBBER SPRING

TRUMPET

FIG 5:21   Mini rubber suspension (front)

6 For competition use it is worth having the torsion bars crack tested to ensure complete reliability (guaranteed crack tested bars are available from Leyland ST for the Marina).

Mini rubber spring suspension:

Early Minis and the later Mk 2 models use the Moulton rubber cone suspension often referred to as the 'dry' type.

At the front the suspension consists of upper and lower arms with the inner ends pivoted on the subframe and the outer ends secured by ball joints to a swivel hub. The hollow rubber springs in the subframe turrets are acted on by the upper arms via hollow trumpet-like tubes. Telescopic dampers are fitted to control wheel patter.

At the rear the suspension is by trailing arm mounted on the rear subframe and acting on a horizontal hollow spring unit via a slightly longer tube than at the front.

Lowering is achieved by sawing pieces off the lower or thinner end of the tubes. Removing $\frac{1}{2}$ inch from the front tube produces $1\frac{1}{2}$ inch lowering; at the rear only $\frac{3}{10}$ inch has to be cut to produce the same result. Shorter telescopic dampers have to be fitted. An alternative modification is to fit the adjustable Hi-Lo units specially designed for the Mini.

The rear suspension tubes can be removed by careful levering with the wheels free. The front units require a special tool available from tool hire specialists. It is used from inside the engine compartment.

British Leyland Hydrolastic suspension:

The hollow rubber cone suspension developed by Alex Moulton for the early Minis was taken one stage further in the Hydrolastic suspension system. The rubber springs were jacketed in a steel case and filled with an alcohol/water fluid (with anti-corrosive additives). Front and rear units (called displacers) on the same side were connected by high pressure tubing. The result was that the fluid displaced from the front unit when a wheel hit a bump pressurised the rear unit and raised the rear of the car, thus counteracting any pitching tendency. The concept was not very successful on the tiny Mini, so eventually the suspension was changed back to solid rubber. However the pitch control on larger cars remained a significant advantage and was used on all transverse engined cars (1100/1300 to Wolseley 2200) until the introduction of the Allegro and more recently the Morris/Austin/Wolseley 18/22.

Lowering the Hydrolastic system is simply a matter of bleeding out some of the fluid. Pressures are very high in the system – up to 250 lb/sq inch – so the job should be carried out at a garage with a special pressurising unit. Specially stiffened front displacers are available for Minis. The Mini and the rest of the cars in the Hydrolastic group should also be fitted with progressive bump stops (Leyland ST or Aeon can supply them) and additional telescopic dampers.

Hydragas suspension:

A further development of the Hydrolastic suspension system is the Hydragas type fitted to the Allegro and 18/22 range of cars. It has a nitrogen gas sphere on each displacer unit taking the place of the rubber 'spring'. Tuning modifications are not yet available from Leyland ST – on no account should depressurisation of the system be attempted at home.

Hydropneumatique suspension:

Nitrogen gas is also used in the Citroen Hydropneumatique self-levelling suspension system which is pressurised hydraulically by an engine driven pump. Citroen specialists can advise on the possibility of modifying this system by fitting different valves.

5:7 Other suspension aids

Anti-roll bars:

The relationship between the stiffness of the front and rear suspensions affects the balance of the car a great deal as explained previously. Many of the measures discussed so far have had the subsidiary benefit of improving the stiffness of one or both ends of the car. However, by far the greatest improvement in roll stiffness at one end can be gained by fitting an anti-roll bar or an extra anti-roll bar to supplement the standard type.

This is a modification which is second only to uprated shock absorbers as an immediate, almost foolproof and relatively cheap improvement to make.

The great advantage of an anti-roll bar is that it does very little to detract from the car's ride but it does improve the handling considerably. It can be used to correct undue understeer or to induce understeer in cars with inherent oversteer characteristics.

A front anti-roll bar will increase understeer (or reduce oversteer) and a rear anti-roll bar will counteract understeer or emphasise oversteering characteristics. A rear anti-roll bar reduces the understeer of the Toledo to give more Dolomite-like handling. Sprites and Midgets on the other hand do not have enough inherent understeer for comfort and a front anti-roll bar greatly improves the normal oversteering tendency of this car.

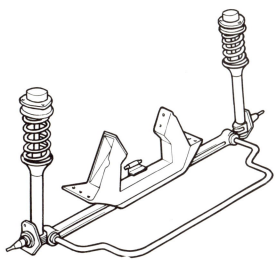

FIG 5:22 Front suspension with anti-roll bar

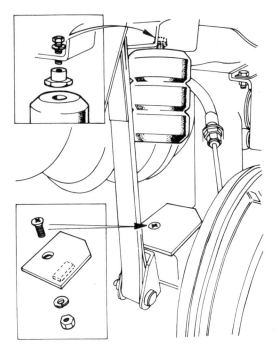

FIG 5:23   Aeon Easiride progressive bump stop

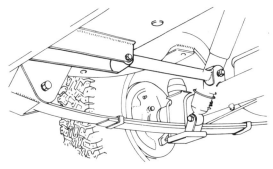

FIG 5:24   Anti-tramp bar

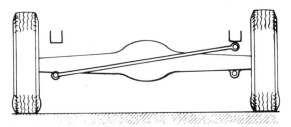

FIG 5:25   Panhard rod

Tuning experts have established the types of anti-roll bar and where they should be placed by long experience of the characteristics of various cars. It is unwise to depart from the action they advise. Tuning shops will supply various grades of anti-roll bar as additional equipment to bolt on to the existing component or as a new fitment. Kits of parts for fitting are usually sold with the bar.

Spring assisters and progressive bump stops:

In cases where it is not possible to stiffen springing or where lowering will cause frequent contact with the car's bump stops, spring assisters or progressive bump stops have to be used. There are various types on the market. A short selection is listed here:

1 Aeon Easiride. Hollow rubber springs are provided in a kit to replace or supplement the car's standard bump stops. These are progressive bump stops that provide additional springing through the last few inches of the suspension's travel (see **FIG 5:23**).

2 Auto balans. Inflatable rubber balls inserted inside coil springs or like Aeon units, over leaf springs. The balls are inflated to between 6 and 16 lb/sq inch and provide provide progressive stiffening through the last few inches of the spring travel. Kits with the correct number of inflatable balls for a particular car are available.

3 McKinney leaf spring assister. The McKinney device is simply a short extra leaf clamped to the underside of the standard leaf spring.

4 Adjustable dampers with auxiliary spring. Built like small MacPherson strut units, damper and auxiliary spring units like the Woodhead Loadmasters and the Spax Even Keel provide a way of uprating suspension and damping at the same time. They are particularly useful for those types of suspension where it is not possible to provide additional stiffness by other means – for example, Hydrolastic, torsion bar.

5 Anti-tramp bars. The tendency for rear leaf springs to tramp and patter due to poor axle location and wind up of the springs can be counteracted by the fitting of one or two anti-tramp bars located between the lower spring mounting and the body. These devices are widely available in specific design for many cars. Frequently it will be necessary to strengthen the area of the boot floor where the bar is fitted. A reinforcing plate should be welded in place and the mounting bracket welded or bolted to it (see **FIG 5:24**).

6 Panhard rod. A Panhard rod provides lateral axle location. It is fitted between a suitably reinforced mounting on the underside of the car body and a bracket welded to the axle casing. The rod should be strung in line with the axle and, although it may be necessary to provide a suitable turret on which to make the body mounting, the rod should be as near horizontal as possible to minimise bump steer and wheel lifting problems (see **FIG 5:25**).

CHAPTER 6

Wheels, tyres and brakes

6:1 Wheels

6:2 Tyres

6:3 Brakes

A complete set of reactions that have to be understood by the tuner for performance or economy occur at the point where the power is transmitted to the road – the wheels, tyres and braking system. Wheels and tyres affect handling, roadholding and performance while the brakes, and the way in which they are used, are vital to the safety of the car. They also contribute considerably to the speed attainable in competitions. The combination of the three systems presents the tuner with an infinite variety of adjustments within which it is exceedingly easy to make expensive mistakes, so each section in this chapter is prefaced by an introduction explaining some of the terms used to describe components and characteristics.

6:1 Wheels

Basic models in the ranges of mass produced cars are usually fitted with the narrowest wheel rim that is consistent with an acceptable rate of tyre wear and reasonable handling and roadholding ability. So using a wider rimmed wheel and tyres to match will, within limits, improve one or all of these characteristics. But before it is possible to discuss the best wheel type to use on a car it is necessary to explore the design of the wheel.

Standard car wheels are cheaply produced items, stamped, pressed and welded from steel plate. They consist of a centre disc stamped with slots for air cooling, weight reducing and styling and holes for the wheel studs. This disc is welded into the rim. The rim design determines the nominal fitting diameter and the effective rim width which is the distance between the tyre retaining shoulders. Central or to one side in the rim is the well which is designed to allow the tyre bead to be placed over the wheel during tyre fitting (see **FIG 6:1**).

A system of nomenclature is used to determine three critical parameters of the wheel design, the rim width, the design of rim and the nominal fitting diameter. A typical wheel would be classed as $5\frac{1}{2}$J × 13. This code signifies a $5\frac{1}{2}$ inch rim width and a 13 inch diameter, the kind of wheel that fits many small popular cars like the Marina and Cortina. The 'J' signifies the shape of the rim and the well depth – B, C, D, J, JK, and K designs are used.

This code does not define the wheel absolutely – there are two other main variables. First is the number of stud holes and the pitch circle diameter of the holes. Second is the inset of the centre disc from the rim centre line. It follows therefore that although a Marina and Cortina may share the same size of wheel the standard wheels on each model may be quite different.

In the later section on tyres the ways in which roadholding and handling improvements are made are discussed. At this stage it suffices to say that for performance cars – but not for economy – improvement will always mean a change to wider wheels. There are a set of rules which must be observed in planning this change:

1 Fouling. Wider wheels and correspondingly wider tyres are likely to create considerable fouling problems, which can lead to dangerous tyre sidewall damage. Standard wheel arches may be too narrow to accept the larger tyre, particularly with lowered suspension. There must be at least $1\frac{1}{2}$ inch clearance all round between tyre and arch or body at full suspension compression. Fouling on the body on lock is also a problem when fitting larger wheels at the front. Brake and suspension pipes can be damaged too.

2 Wheel inset. To minimise wheel bearing wear the centre line of the wheel rim should run through the wheel

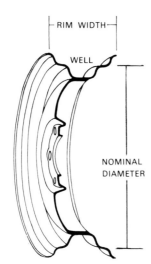

FIG 6:1   Cross-section of steel wheel

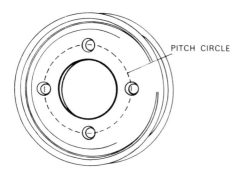

FIG 6:2   Defining the position of stud holes

bearing (or as near to it as possible) which means that with various types of brake and wheel hub the manufacturer varies the wheel disc inset from this line. A certain amount of extra stress on the bearing can be accepted in a performance car but it is not advisable to vary the rim centre line much more than one inch from the standard wheel location. Wheel inset does of course have an effect on the clearance. Very wide wheels may have to have an offset disc so they can be accommodated in any kind of wheel arch (see wheel arch extension, **Chapter 7**). Once again the effect of this shift of forces on the wheel bearings must be taken into account.

But perhaps the greatest danger is that the wider wheel will begin to affect the car's steering characteristics. Most cars have front wheel 'toe-in' so that as the car picks up speed the wheels are forced back taking up the play in the steering system until at speed they are parallel. Wider wheels increase the drag so at speed the wheels may begin to 'toe-out' – the result is that the steering will begin to feel very vague. The additional scrubbing action of the tyres reduces their useful life.

3 Ride. This same wheel factor, the inset/offset, also affects the car's ride. Because the tyre is farther out the leverage effect of the suspension is altered in such a way that the dampers actually telescope less and therefore they are less effective. The bigger tyres do not conform

to road surface irregularities quite so well as narrower treads nor will they compress so much; the result is a harsher ride. The third ride problem is the increase in unsprung weight. The use of alloys in the construction of wider wheels gives two advantages – greater strength and lower weight compared to steel wheels of the same size. The wheel weight forms part of the car's unsprung weight and any increase in this weight, which in effect dictates the immediate response of the car to bumps, causes more violent crashing up and down.

4 Tyre grip in the wet. Although in the dry the wider tyre gives greater grip because of the improved interference between the road surface and the tyre (a larger contact patch) the pressure exerted by the tyre on the road is in fact smaller because the same weight is spread over a larger area. It is this pressure of tyre on the road that aids the ejection of the water film from the contact patch and maintains tyre adhesion. A lower pressure reduces the ability to squeeze out the water and thus a wider tyred car may lose adhesion earlier than a standard tyred car in the wet. This loss of adhesion is due to aquaplaning.

Wheel construction:

Wheel widening specialists sell or make wheels of four basic types. The widened steel wheel is a conventional wheel cut around the rim with a band of the required width welded into place (see **FIG 6:3**).

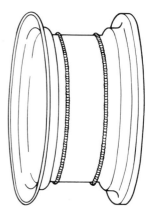

FIG 6:3   Steel wheel with welded-in band

As mentioned previously these wheels are quite heavy and the welds can be porous so that the air leaks out. The solution is to use an inner tube, thus adding further cost and weight to the wheel assembly. Unfortunately this unsatisfactory assembly is still cheaper than the next type of wheel which is custom made from steel. Various styles are available.

Reducing the wheel weight to solve unsprung weight problems involves using aluminium alloy wheels and, again, there is a very wide range to choose from at the speed shop or wheel specialist. The lightest wheels of all are the magnesium alloy types like the Minilite. These wheels should be considered as competition modifications only. In road use they can corrode and become unsafe quite apart from the expense of equipping a car with them.

In fact alloy wheels in general require a great deal of care that is usually unjustifiable in a car used for normal motoring. Kerb scraping can remove enough of the soft metal to cause rim damage and affect balance and great care has to be taken over the tightening of the retaining nuts. The correct procedure is to tighten the wheel nuts to the correct torque specified by the wheel manufacturers and run the car for up to 100 miles. The nuts should then be retightened to specified torque and run a further 400 miles. A third tightening will probably be needed. The procedure is necessary because the alloy is softer than steel and the wheel nut holes deform slightly under load. Since thieves are attracted by these expensive items a set of locking wheel nuts is a good investment.

Wheel spacers:

In the early days of the performance equipment boom one of the best selling accessories was the wheel spacer used to increase a vehicle's wheel-to-wheel width. The spacers were sold with long wheel studs and the combination was potentially very dangerous. Hub wear was so accelerated that hubs would overheat and shear under the additional loads these spacers created. The problem was compounded by hanging wide wheels on to the end of the spacers.

Now, opinion is totally against the use of spacers to increase track width. It may be permissible to use them in certain instances to preserve the correct ratio between the centre line of the wheel and the bearing, that is to alter the wheel inset. Where this necessitates the use of extended studs, obtain the advice of the car or wheel manufacturer on the effects this will have on the wheel hub. It may be necessary to obtain competition wheel hubs.

A cheap source of wide wheels:

Like most other aspects of tuning the answer to cheaper wide wheels may be found further up the car manufacturer's model range. High performance versions of basic cars usually have wider wheels and as these are standard original equipment items they are often cheaper than the equivalent accessory wheel. The other advantage of this approach is that the wheels usually fit with no trouble. The only problems that occur are that the wheel arch clearances may not be sufficient in the basic model or that front disc brake assemblies on a sports model have been designed to give more room for wide wheels than a basic drum brake set-up. These points are easily checked on a showroom model.

Wheel attitude:

The stability of the wheel at speed and the car's understeer or oversteer characteristics are affected by the camber of the wheel; in effect its attitude to the road. Positive camber is the name of the condition in which the wheel leans outward at the top, and the opposite condition is negative camber in which the wheel leans inward at the top (see **FIG 6:5**).

Camber changes are for the skilled performance specialist only, as for most cars the techniques involved require special workshop facilities. In addition this is not one of those modifications that can be experimented

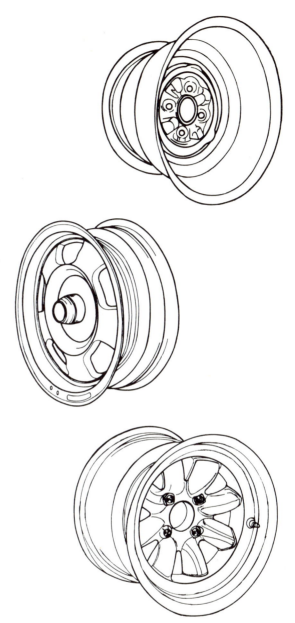

FIG 6:4　Wide rim steel and alloy wheels

with cheaply. In some instances, however, manufacturers' competitions departments can supply camber change kits that are not too difficult to fit. The Vauxhall Viva is an example: Dealer Team Vauxhall can supply shortened top links for the front suspension.

The general rules for a change of camber are that understeer can be induced by giving negative camber to the rear wheels and retaining normal camber at the front, or increasing positive camber at the front and retaining normal camber at the rear. Reducing understeer or inducing oversteer can be performed by increasing negative camber at the front.

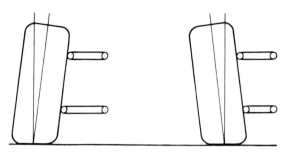

FIG 6:5 Negative camber (left), positive camber (right)

6:2 Tyres

Like wheels, tyres are also size coded. A typical tyre might be designated 5.60 × 14. This means that the widest part of the tyre measures 5.60 inch, sidewall to sidewall, and that it will fit a 14 inch wheel. This Imperial measurement for tyres is now being used less and less – the Metric equivalent to the example above is 165 × 14.

Tyres are also speed rated by a letter inserted into the size code, for example 165S × 14. For cross-ply tyres the following table gives the tyre use limitations:

Rim diameter	No marking S	CODE H	V
10 inch	75 mph	95 mph	110 mph	over 113 mph
12 inch	85 mph	100 mph	115 mph	over 115 mph
over 13 inch	95 mph	110 mph	125 mph	over 125 mph

The maximum speed limits for radial tyres are higher and a designation Radial or code R (for example 165SR × 14) will be given to distinguish them from cross-plies. The limits apply to all tyre diameters and are as follows:

SR	HR	VR
113 mph	130 mph	over 130 mph

Since October 1974 tyre manufacturers have graded the quality of their finished products into three groups. These are classified as New, Regraded Quality and Seconds.

New: New tyres without blemishes or moulding defects.

Regraded Quality: New tyres that have minor moulding defects, but speed ratings are not affected. The SR, HR, VR coded markings are retained.

Seconds: New tyres with structural defects which, although not serious, restrict their use for high speeds. These tyres carry no speed codes on the sidewalls and they have the same speed ratings as remould tyres: up to 70 mile/hr (10 inch rim) and 75 mile/hr (over 10 inch rim).

Before October 1974, the classification Remould Quality (known as RQ) embraced a wide range of production defects and for safety reasons the tyres were restricted to the same speeds as remould tyres. Unfortunately, some still retained the high speed markings and on others the markings were erased, which left the motoring public uncertain as to their use. The vagueness of the whole situation prompted a reclassification.

However, confusion may still arise if the abbreviation RQ, used by the trade for the obsolete Remould Quality, is tagged to the later classification Regraded Quality.

Real remoulds, tyres which have been remanufactured from the belting upwards, should bear a British Standard Institution marking and the name of a manufacturer. The speed limits for uncoded tyres apply and, although they are fit for normal motoring, they are not considered suitable for rally or frequent fast road use.

Another tyre factor that is increasingly mentioned is the aspect ratio. Tyres with the same width and height of cross section have a 100 per cent aspect ratio – some early covers were as square as this. Old fashioned cross-ply tyres have an aspect ratio of 95 per cent. Their modern counterparts and many radial tyres are designed with an 82 per cent aspect ratio. The latest tyres have a wider tread than previous types with an aspect ratio of 70 per cent. The first of these series were developed for the Jaguar XJ6 and they have been widely used in sports applications ever since.

The lower the aspect ratio the shorter the rolling radius of the tyre. Thus a change from an 82 per cent tyre to a 70 per cent lowers the car's final drive ratio. The ramifications of this lower ratio are discussed in **Chapter 4** but the effect is a gain in performance at the expense of a loss in economy which can only be set right by a change in rear axle ratio or the use of much wider tyres and wheels. Thus it is disadvantageous to change to these modern tyres from an economy point of view. The other factor involved in this change is the considerable inaccuracy it introduces in the odometer and speedometer. Instrument manufacturers should be able to advise on the change in drive gear or speedometer head that will be necessary to overcome this problem.

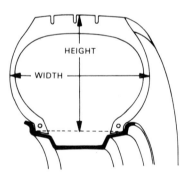

FIG 6:6 Aspect ratio: height divided by width

The advantages and disadvantages of wider tyres:

The only direct advantages of fitting wider wheels and tyres are a gain in roadholding in certain conditions and a gain in traction for a standing start. The conditions are in the dry and on good surfaces.

The road holding advantage is, however, sufficient to overcome some of the disadvantages especially if the size change is relatively small. Some of the disadvantages to be overcome have been discussed previously in the section on wheels. It is important to bear in mind that a one inch change in wheel rim width will generally give rise to few of the problems discussed there. It is probable that for a sensible increase of this kind the only appreciable disadvantage would be slight deterioration of ride.

Tyre tuning for economy:

There are no economic advantages to be gained from changing wheel width or tyre size. However, motorists with cross-ply tyres should seriously consider a change to radials when the next tyre swop is due. This is because radial tyres offer a lower rolling resistance to the car's progress than equivalent cross-ply covers. The difference could mean as much as a constant 2 to 4 per cent petrol economy.

FIG 6:7   Construction of a cross-ply tyre

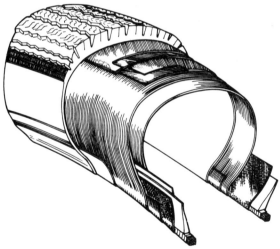

FIG 6:8   Construction of a radial tyre

The next largest economy can be made by maintaining the tyres at the manufacturer's highest specified pressure. This pressure is usually given in the car handbook as an increase of 2 lb/inch over the normal pressure (the same as for high speed or heavily laden conditions). A 2-3 lb/inch increase over the normal pressure in all motoring conditions is permissible for the simple reason that car manufacturers usually give a deliberately slightly low tyre pressure because it softens the ride.

The primary advantage of the higher pressure is that once again the rolling resistance is reduced. But constant attention to the pressure of inflation will also pay dividends in increased tyre life.

For correct tyre maintenance a good quality tyre pressure gauge is essential. Never trust the air-line pressure gauge on a garage forecourt. The conditions of abuse they are subjected to lead to frequent gross inaccuracy.

Rubber mix, tread pattern and reinforcement:

Tyre manufacturers vary the rubber mix to suit the conditions for which the tyre is designed. For most normal motoring applications the manufacturers use a mix of rubber that provides the best grip with the longest life. These two factors are mutually exclusive – the grippier the tyre the more it will wear. Tyres also vary according to the preferences of the home market in the country of origin. For example, Japanese tyres have a noticeably harder type of rubber than equivalent European tyres and certain of them have a marked lack of grip in the wet for this very reason. These harder tyres do, however, last much longer than softer European types.

Every manufacturer's tread patterns vary to a large degree. Modern patterns are designed to provide a compromise between water explusion, traction in slippery conditions and, once again, tyre life.

Tyres are reinforced by fabric made from man-made fibres like rayon. Some designs are additionally braced by woven steel mats.

Matching tyres to wheels:

Tyres are manufactured in such a way that strictly they are ideally suited to only one rim size. These matched sizes are as follows:

Rim size (inch)	Tyre size (mm)
$4\frac{1}{2}$J	145
5J	155
$5\frac{1}{2}$J	165-175
6J	175-180

There are exceptions to every rule: consult the tyre manufacturer about the possibility of fitting a tyre one size larger than the recommended maximum. The potential danger in doing this is that the extra sidewall and tread deflection that can take place on a rim that is too narrow may enable the tyre to roll off the wheel. (This is a particular problem with oversize low profile tyres). Centre tread wear will also be accelerated.

Tyre care:

New tyres should be run in. Avoid harsh acceleration and braking for the first 100 miles as this could provoke excessive heat rises in the tyre until it has flexed enough to loosen up.

Tyre and wheel should be balanced when new and every 5000 miles, or when excessive wheel vibration is felt. Static balancing is carried out on a special instrument at the tyre dealer or garage. There is really no home substitute for the expertise of the tyre fitter in sorting out the balance problems for sports or performance cars.

Should the static method prove inadequate it should be possible to find a dealer offering dynamic balancing

93

on a rotating machine. The ultimate type of balancing is carried out with the wheel on the car. This process, carried out with a miniature rolling road, takes into account any imbalance in the hub and brake drum.

6:3 Brakes

Much of this book has been devoted to making the car go faster, albeit in greater safety. The aim of the tuner's attention to the braking system is the exact opposite of this – to make the car stop faster. To this observation must be added the rider that it is also necessary to make the car stop in a straight line.

The action of hydraulic brakes:

Brake pressure is applied to the friction pads of the brakes through a pedal actuated hydraulic system. The pads, pressed hard against the drums or discs, have to overcome the turning force exerted on the wheels through the tyres which is a reaction between the car's momentum and the friction or adhesion of the road surface.

This is a relatively simple picture of the operation of the brakes which is true for most normal situations and for most standard saloons. Faster performance cars are used in slightly different ways to the standard car. In competition the brakes will be pushed much harder than in normal road situations with frequent and severe braking from speed. Far harsher brake usage on normal roads is also likely to occur if the car's performance is to add to the pleasure of driving and the car's flexibility is utilised to make the most of overtaking and other traffic situations.

Performance tuned cars accentuate the weaknesses in braking systems in situations in which there is less room for error. The following characteristics are common to the brakes of most cars and it is not just tuned cars that could benefit from the maintenance and modifications suggested.

Wheel lock-up:

Locking wheels describes what happens when the wheel comes to a halt before the car and it occurs as a result of a weakening of the tyre's grip on the road. There are a variety of reasons for the loss of grip – oil, ice and water on the road are common causes of lock on one or all wheels and these are to some extent unavoidable. Poor tyre tread is another cause and that can be avoided by regular inspection and replacement. But the major lock-up problem in braking is caused by weight transfer.

Most cars are front heavy – fast braking shifts even more of the car's weight on to the front wheels and lightens the load on the rear wheels, to the point where tyre grip is lost and the rear wheels lock up. The locked wheels are poorly braked and directionally unstable; they attempt to overtake the front wheels and the car enters a spin unless controlled.

Various ways have been attempted to control the weight transfer characteristic. One of the most common is the pressure limiting valve (sometimes called the load transfer valve). This unit is installed in the brake pipe between front and rear brakes. When pressure is applied to the hydraulic system it is transmitted equally to front and rear brakes until a certain pressure is reached. At this

point the valve shuts off (maintaining the rear brake line pressure) and any subsequent pressure rise is transmitted solely to the front brakes. The level is preset by the manufacturers to take account of most normal and emergency braking situations on road surfaces. In abnormal braking conditions and on broken surfaces (when rallying) the balance is destroyed.

Cars with dual circuit brakes may have a means of adjustment whereby the amount of pedal travel applied to each cylinder is varied. Other cars rely on different sized wheel cylinders in the front and rear.

Grabbing brakes:

Out-of-roundness of the brake drums, irregular thickness of the discs or patchy wear of the friction materials can cause brake grabbing in which one or more wheels react more savagely to braking than the others.

Loss of braking on one wheel:

Wheel cylinder sticking or a hydraulic failure, brake fluid or oil contamination of the pads or linings and scoring of the discs or drums can cause diminished braking in one wheel or even total loss of braking effort.

Brake fade:

Rapid repeated use of certain types of linings causes a temperature related loss of friction properties that has been termed brake fade. Generally, the softer the brake lining material, the lower the pressure required to achieve satisfactory braking and the lower the resistance to fade. Harder linings that resist fade well require higher pedal pressures. Manufacturers choose the car's standard brake lining material to be a compromise between fade resistance and pedal pressure. Fade can also be unnecessarily prolonged by poor cooling of the brakes. Under normal circumstances natural cooling will restore brake operation within a few minutes.

Brake maintenance for performance and economy:

Brakes are vital to car safety and regular maintenance can prevent the onset of many of the above faults as well as save money.

Drum brakes:

Fit new shoes if linings are worn to within $\frac{1}{16}$ inch of the rivets (or $\frac{1}{16}$ inch thick if they are bonded shoes). Always fit new shoes to both wheels on the same axle. Details of the dismantling, rebuilding and adjusting of individual types of brakes are given in the car's workshop manual. General points to watch are:

1 Note shoe positions so replacements are fitted correctly.

2 Wire or otherwise secure wheel cylinders so that residual pressure does not force pistons out with consequent loss of fluid.

3 When cylinders are removed block off flexible hydraulic hoses with proper clamps and plug the refill cap of the master cylinder with a rubber bung or by screwing the cap down onto a small piece of polythene.

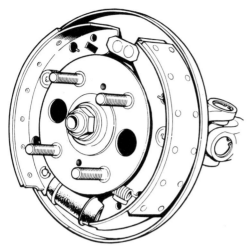

FIG 6:9 Drum brake assembly with drum removed

4 Brake parts can be cleaned with a methylated spirit soaked rag.

5 Lubricate pivot points with high temperature brake grease.

6 It is important to adjust the brakes correctly to achieve the shortest possible brake pedal travel without binding of the brakes, that is, without producing drag that would affect fuel consumption. Test for brake binding by firstly listening for the sound of linings scraping on the drums and, secondly choosing a quiet stretch of road and driving for about $1\frac{1}{2}$ miles **without using the brakes**; coast to rest and feel the drums for the warmth that will indicate drag.

7 New brake shoes should be bedded in by making about a dozen full emergency stops from 20 mile/hr and then readjust. After about 200 miles new pads should be examined and brake dust should be brushed out of the drums.

8 Examine brake pads every 3000 miles and clean down the assembly, regreasing if necessary. Pay particular regard to the condition of the drums and note that a scored contact area will mean drum replacement. Skimming is not a satisfactory solution as it reduces capacity for adjustment.

Disc brakes:

Most types of disc brakes can be overhauled by the home mechanic though a few rear disc brake units need special tools for pad release and removal and may have to be dealt with by a garage. General rules are as follows :

1 Generally replace pads when $\frac{1}{8}$ inch of material remains although some manufacturers recommend replacement sooner.

2 Some types of disc brakes, the swinging caliper design, have pads with a wedge shape. These should be replaced when worn parallel with the disc.

3 Inspect disc for scoring. If there are any signs of this the disc must be replaced.

4 Clean rim of rust off the disc with emery paper but if there is rust on the braking surface this is a sign of a seized caliper.

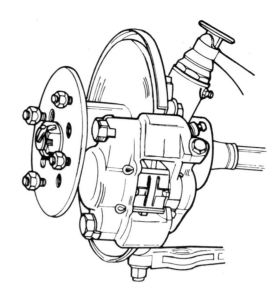

FIG 6:10 Fixed caliper disc brake assembly

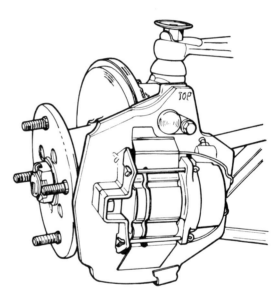

FIG 6:11 Swing caliper disc brake assembly

5 Rubber dust covers should be inspected for perishing. Look under the cover for fluid leaks.

6 Discs should be checked for distortion **before** fitting new pads. Use a feeler gauge between the brake housing and the spinning disc. No more than .005 inch distortion is permissible.

7 Closely observe instructions and arrows on pads to get correct fitment. Anti-squeal shims (which should be lubricated with high temperature brake grease) are also marked with arrows for correct location.

8 Adjustment of the handbrake caliper should be carried out after pad renewal and a clearance of .003 inch should be aimed for. However, many modern units are self-adjusting.

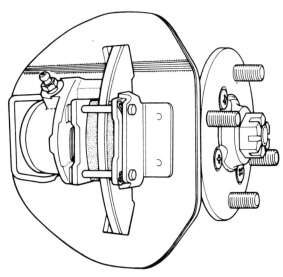

FIG 6:12 Sliding yoke disc brake assembly

Hydraulic system maintenance:

Brake fluid should be renewed every 10,000 miles. Hoses should be renewed every 30,000 miles. Inspect the entire hydraulic system for leakages every 3000 miles and pay particular attention to the possibilities of chafing between flexible hydraulic pipes and components of the steering or suspension system. Individual operations will be described in detail in the relevant workshop manual:

1 When renewing parts of the hydraulic system always buy **all** new parts including such seemingly insignificant bits as washers.

2 Metal pipes should be replaced with parts of the correct grade but it is worth considering the use of a modern material like Kunifer 10. Parts specialists will provide a kit or make up pipes to the correct length and shape.

3 Master and wheel cylinders that leak should, ideally, be replaced by complete new units. Overhaul kits are often available but there may be internal wear that will render the unit unserviceable in a very short time.

4 There is no substitute for thorough bleeding of the braking system after a service and during fluid renewal when all old, dirty fluid should be forced out. A brake bleeding aid like Gunson's Eezibleed is a valuable time saving addition to the tool kit for this work. Start bleeding at the nipple furthest from the master cylinder and work round the car until no air bubbles can be seen entrained in the fluid bleed (front disc systems should be bled first – on dual circuits the rear circuit should be bled first).

Handbrake maintenance:

Rear brake drag is often the product of poor handbrake adjustment. Cable runs are notorious for clogging with dirt, closure through stone damage and even being fouled by underseal. Handbrake adjustment should be carried out according to the individual workshop manual. General rules are:

1 Pay particular attention to the entire run of the hand-

brake cable watching out for sluggish cable return movement and battered guide channels. Repair or clean where necessary.

2 Lubricate any pivots in the system frequently, either with grease or with heavy oil. Graphite lubricants can be used on cables running in tubes or channels.

3 Dissolve off underseal that is affecting cable or lever movement with petrol.

4 Maintain as accurate side to side handbrake balance as possible when adjusting and ensure that the shoe return spring mechanism is operating as freely as possible.

5 Replace tired or corroded cables immediately.

Brake Modifications:

Brake linings:

Harder linings are the first brake modification an aspiring performance tuner turns to. These linings, like Ferodo VG 95 for shoes and DS11 for pads, have greater resistance to fade than the materials fitted to standard cars. The penalty is the large increase in pedal pressure that will be required to obtain anything like reasonable brake performance from the shoes or pads. For a road car a servo will almost certainly have to be fitted to maintain comfort, ease of operation and safety. For cars that already have servos the pedal pressure increase may be acceptable but consult the brake manufacturer for performance details. There are several rules that must be observed in using these pad materials.

1 Shoe linings can be bought separately with a set of extra strong rivets. Unless expert knowledge and help is available don't attempt to line brake shoes. Obtain bonded or ready riveted shoes.

2 Observe the bedding in procedure described for new

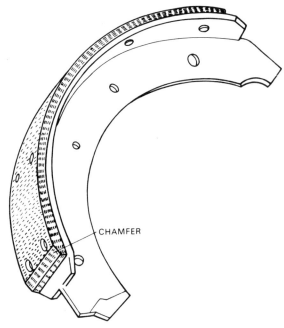

FIG 6:13 Brake lining with leading edge chamfered

shoes (see Drum brakes, 7) otherwise an emergency could find the car with no effective means of stopping.

3 Fit the shoes all round. Fitting to one axle only has drastic affects on the car's braking capabilities that can only be explored by cautious experiment and only turns out to be advantageous in a very few circumstances.

Slightly smoother and more rapid bedding in of shoe linings and a solution for occasional problems of brake judder can be achieved by chamfering the leading edge of the lining material. Chamfer to a depth of about $\frac{1}{8}$ inch with a coarse file and finish smooth with emery paper.

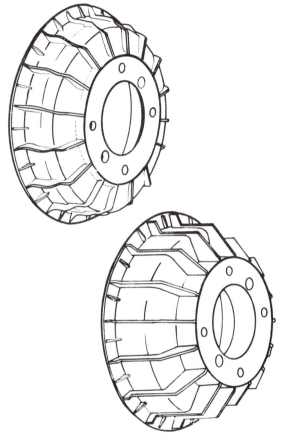

FIG 6:14   Minifin light alloy brake drums

Uprated drums:

One of the major factors in the promotion of fade is the temperature characteristics of the brake and wheel combination. Standard cars used in a conventional manner do not need the close attention to the cooling of the drums that makes the difference between average and good stopping ability in repeated brake use that is demanded of a performance car. However if fade problems occur and a change to harder linings is not desirable it is possible to purchase a special type of brake drum for some models, giving extra cooling ability. The drum is cast in alloy with good heat conducting properties. The casting is finned and has an iron insert in the shoe contact area. One make is the Minifin which is available for a number of popular cars.

Brake swops and disc brake conversions:

Swopping the disc brakes from a sports or larger engined model to the basic model is not the easy process it sounds. The manufacturer frequently makes various changes to the front suspension, steering and track rod components of the wheel system that are difficult to spot without expert knowledge. However, it is possible to do this on certain models of car and dealers and main agents may be able to advise which models have exchangeable systems. Larger rear drums may be obtained for older cars from a scrap yard. Look for a similar axle type and find out if the combination will fit.

Alternatively many manufacturers sell disc brake conversion kits through their competitions department. The kits are easily fitted and take into account the variations in layout that have occurred on differently equipped models.

Front to rear balance modifications:

As mentioned above the tendency for early lock-up of the rear wheels can be overcome in a few ways. One of the easiest modifications to carry out, on some cars, is to obtain smaller sized rear wheel cylinders. These may be found on less powerful models in a range or on a completely different model altogether. Consult the brake system manufacturers to find part numbers. Do not use scrapyard components for this safety modification.

It is usually only necessary to go one size smaller to obtain a very large reduction in rear braking effort. The opposite modification to this, suitable for a rally car in which a tail-out configuration of cornering may be considered desirable, is to increase the rear braking effort. This can be done by fitting a size larger (say an extra $\frac{1}{8}$ inch diameter) wheel cylinder.

A more advanced modification is to tackle the transfer valve (pressure limiting valve) itself:

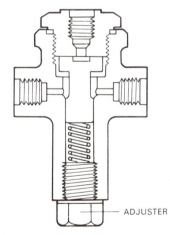

FIG 6:15   Modified brake limiting valve

1 Remove the valve from the system and unscrew plug at input end of valve.
2 Remove plunger and spring taking care not to damage seals.
3 Drill through normally blanked spring recess end.
4 Internally or externally thread this arm of the valve

assembly to take a bolt or screw down cap that will bear with varying pressure on the spring (see **FIG 6:15**).

5 Reassemble valve, refit in system and bleed out. Extensive testing should be carried out to find the optimum setting of the adjustable bolt or cap. Once adjusted, protect the valve unit with a bolted-on conduit box; alternatively cut an access hole to the unit through the car floor and fashion a permanent cover for it.

Safety braking hydraulic system modification:

A device called the Safety Braker can be obtained that deliberately introduces an adjustable element of sponginess into the hydraulic system. This enables pads or shoes to ride over irregularities in the drum or the disc without destroying brake performance. It is a form of shock absorber which is installed directly in a convenient place in the hydraulic system – convenient for driver adjustment if possible.

Fitting a brake 'booster' or servo unit:

A servo does not improve braking performance – it simply makes it easier for the driver to achieve maximum braking effort. Useful in overcoming the high pedal pressures found when harder linings are fitted, servos can be obtained in kit form with all the parts necessary for fitting to a wide range of cars. An example of this type of kit is the Girling Powerstop.

Before fitting a servo ensure that all components of the braking system are in top class condition – it is wise to carry out a thorough overhaul and replacement of all hydraulic components if the car has over 20,000 miles on the clock. The major steps of the fitting procedure are:

1 Find a suitable location for the unit on the engine compartment bulkhead as close as possible to the master cylinder. The unit should be mounted horizontally – never placed on or touching the engine.

2 Either drill and tap the inlet manifold and fit a vacuum connection or fit a T-piece to the existing vacuum gauge connection or use a carburetter flange spacer with a vacuum take-off.

3 Connect the pipe from the master cylinder to the unit using the existing tube or the new length supplied.

4 Run new piping to the connection point vacated by the master cylinder connection.

5 Connect up vacuum tube.

6 Bleed the braking system thoroughly (it may be possible to prevent complete loss of system fluid by delaying connection to the rest of the system until after the servo itself has been bled).

CHAPTER 7

Bodywork, accessories and instruments

7 : 1 Fitting glass fibre parts
7 : 2 Metal reinforcement
7 : 3 Customising the interior

7 : 4 Anti-theft devices
7 : 5 Auxiliary lighting

7 : 1 Fitting glass fibre parts

Wheel flares :

Wheel arch flares are absolutely necessary for cars on which wide wheels protrude beyond the standard wheel arches (see **Chapter 8**). The flares (often called wheel spats) can be purchased from accessory shops or by mail order from bodywork specialists. A few are made in metal but the majority are plastics or glass fibre items. Follow this procedure :

1 Offer the wheel spats up to the car and mark around them.
2 Measure the flange on the flares and scribe a second line on the body where the existing wheel arches are to be cut.
3 Cut out the marked area with a power jigsaw or a Monodex cutter. At the rear it may also be necessary to remove a section of an inner arch.
4 Clean down the cut edge ensuring that the surface behind the jointing flange is free from dirt and underseal.
5 Dab down a resin impregnated strip of glass fibre tape along the flange of the flare and apply resin to the joint area of the wheel arch.
6 Tack the flare into place with pop rivets.
7 If an inner rear wheel arch panel is also to be added secure this in place in the same way as the flare and then lay up glass fibre and resin to joint the two parts and strengthen the whole arch area.
8 Sand down the area of the join and use filler to cover the rivets and the sharp metal edge. A scraper cut to the shape of the flare and bodywork can be used to contour the paste down to a minimal camouflaging layer.
9 Carry out preparation for painting.

Air dams and spoilers :

Glass fibre air dams (lips fitted below the car's front apron to alter the aerodynamic characteristics) are available from bodywork specialists. These units, and spoilers which are fitted to the car's boot lid, are specially designed to fit the bodywork contours of various cars. In some instances an air dam is combined with wheel arch flares, e.g. Mini. The stages in fitting these accessories are the same as for the fitting flares although it is not usually necessary to carry out any cutting of the bodywork.

GRP (glass fibre) replacement panels :

A very large industry, making and supplying GRP replacement panels for a wide range of popular cars has sprung up over the last ten years. The fact is that GRP panels cannot offer the same crumple-resistance as the car's original metal bodywork and so a car modified in this way will have a much lower ability to soak up the impact energy in accidents. The result is that more energy is likely to be transmitted to the occupants, hurling them against the nearest piece of the car's interior – thus more injuries can be caused.

It must also be borne in mind that the modern unitary construction car has very few panels which have no body strengthening function. Therefore it is absolutely essential that GRP panels are only used in those applications where no stress occurs. For most cars this limits the application of these panels to the boot and bonnet lids and occasionally the front wings and apron. GRP sills are not advisable for the simple reason that they do not cure the corrosion; a proper sill repair can only be effected by removing all rusted metal and welding in suitable replacement parts.

With these restrictions borne in mind it is possible to use these parts to effect a considerable weight reduction

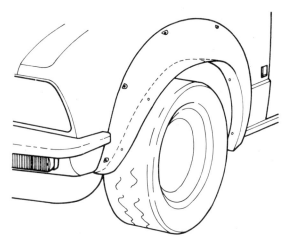

FIG 7:1   Wheel arch flare in position

on a performance car. Boot and bonnet panels are reasonably cheap and much lighter than their metal counterparts.

Further information on the use of GRP, as well as repairs, painting and customising, can be found in another Autocare manual, **Bodywork Maintenance and Repair.**

7:2 Metal reinforcement

If welding techniques are available they can very usefully be used to strengthen a car used for certain types of competition work such as rallying. A great deal of strengthening can be achieved by continuously welding all those body seams which join together stress bearing members of the body. These joints are usually spot welded to a pattern designed by the manufacturers to cut down on the number of welds performed without prejudicing the body strength for normal motoring purposes. In rallies or other hard use this will almost certainly not be strong enough.

Certain additional areas can be given extra strength by welding in reinforcement plates. Most cars with MacPherson strut front suspension will benefit from beefing up the area of the wing or flitch panel to which the strut is mounted (sometimes called a turret). Suitable plates can either be fabricated from mild steel plate or bought as repair items from bodywork specialists.

Under the car, vulnerable parts of the bodywork and the exhaust system can be given protection by small plates welded on to their leading edges. These not only act as strengtheners against road debris thrown up underneath the car but also perform as skids so that, to a certain extent, the treated part will ride over larger obstacles like the centre ridge of a rutted track.

Roll-over bars:

Demanded by the regulations in some forms of autosport and a sensible addition to the safety armoury of a performance modified car, roll-over bars or cages are available to suit most cars. The only draw back is that they take up a good deal of interior space – they don't look very pretty either.

Most types of cage or bar are designed to introduce a rigid hoop of steel over the driver's head or slightly behind it. Some more advanced designs extend the protection offered right to the front of the passenger compartment. The cages or bars are bolted into place on known strong points of the car body.

7:3 Customising the interior

Both the normal family saloon and its uprated performance cousin becomes a good deal more pleasant to drive and distinctive to look at if the interior is also treated to some customising work.

Seating:

Few seats fitted to mass production cars offer the comfort, location or range of adjustment of a special seat chosen from the wide range of designs available to the customer. Bucket-type seats which offer good side or lateral location are an important safety addition in a performance car which is cornered hard.

Never buy a seat without sitting in it to ensure that the lumbar support given to the lower spine is adequate, that the tension of the lower squab is sufficient and that the side support is adequate. Also look out for stout floor mountings which locate the seat at both sides and provide a decent range of adjustment. Seats that have a squab height adjustment facility are particularly adaptable to a comfortable position. Check that it is possible to adjust the head rest (if fitted) to a position where it supports the head at a point level with the eyes. If it is non-adjustable and only reaches to the nape of the neck it could be a killer in a rear end shunt accident as it will not prevent whiplash injury.

Steering wheels and rake adjustment:

The mass produced steering wheel fitted as standard to most normal saloons is not a very distinguished item. It may also be too large for comfort in carrying out the steering juggling demanded in travelling at speed. A more distinctive 'sports' type of steering wheel with a padded plastics rim, perhaps leather covered, is often the first choice of a customiser to personalise the driving position.

There are many makes to choose from – the important points to bear in mind are:

1 Never buy a wheel too small to handle the car in low speed parking situations, especially if a wife or girlfriend is to drive the car as well.

2 Ensure that the wheel will fit the splined or other fitment on the car's steering column.

3 Look for a properly dished or padded boss. Steering wheel injuries in accidents are terrible and can be avoided by selection of a good quality wheel. The performance of wooden rimmed wheels in accidents – they can splinter into sharp pieces – is one reason why they have not been so popular recently.

4 Feel the wheel in all hand positions and ensure that the rim or binding is not too thick for the size of the hand.

Another improvement that can be made to the steering of some cars is the adjustment of the rake of the steering column. The Mini's, in particular, attracted a great deal of attention as a readily adaptable column. Kits are available to lower the wheel considerably but it should be borne in mind that the adjustment requires movement of the steering rack. This must be carried out very carefully to avoid dangerous straining of the steering column spline joint. Recent cars have had the steering column attached to the upper facia mounting by sheer bolts so the adjustment of rake has been made a lot more difficult.

Customising facias:

It becomes particularly important for the driver of a performance tuned car to know a little bit more about what is happening in the engine when on the move. There is an easy way of doing this that can add to the car's facia appearance and that is to fit extra instruments.

Instruments are best installed on a subsidiary instrument panel placed as near as possible to the driver's line of sight and as near to the windscreen level as practicable. It may be thought useful or stimulating to include revision of the layout of the car's original instrumentation at the same time as adding other facia features.

A few general notes on instrument installation connection and wiring will aid the design of an easily read and correctly working instrument panel.

Ammeters must be wired between the main battery supply terminal on the starter solenoid and the charge lead from the dynamo or alternator. It may be convenient to connect it in series between the control box Al terminal and the lead removed from the terminal. As such a joint will carry the full charge/discharge load of the battery generator system it must be made to a very high standard. Soldering is advisable though a heavy duty insulated barrel connector can be used. The wiring should be of a sufficient grade to take the current loads flowing in this circuit – minimum 44/.012 grade wire. It is wise to wrap the joints at the back of the instrument (made with heavy duty Lucar type connectors or large tag connectors) with a thick layer of plastics insulating tape.

A vacuum performance gauge can be used as an aid to economical driving and to diagnose various tuning faults in the engine.

The manifold vacuum connection can be made by drilling and tapping the manifold ensuring that no swarf remains inside the tubes before refitting to the engine. A simpler way is to fit a thick vacuum tapped spacer; the same device can be used for the connection of a brake servo accessory. Lead the plastic vacuum piping away from the hot exhaust manifold and through a hole in the engine bulkhead. Connect it up to the instrument and adjust the damping of the vacuum pipe with the screw-up clamp supplied with the instrument.

There are various styles of tachometer that can be fitted as an accessory instrument. They all work on a similar principle, which is to count the impulses of low tension current supplied to the coil by the switching action of the distributor.

Dial illumination lamps inside instruments can be wired to the existing instrument panel lighting using insulation piercing connectors of the 3M Scotchlok type. If the car has a printed circuit for instrument panel wiring take the dial illumination supply lead from a point on the main lighting circuit between the main light switch and the dipswitch.

As far as possible use existing holes in the engine compartment bulkhead to feed instrumentation and other accessory leads through to the facia. If it is necessary to make additional holes drill carefully to avoid damage to existing wiring and make a hole large enough to accept a protecting grommet.

Site instruments carefully on a redesigned facia or subsidiary instrument panel. For instance vacuum gauges and non-electric oil pressure gauges should be mounted in a position that affords easy kink-free access for the pressure tubes. These two instruments and the ammeter should always be near to the driver's line of sight to make checking easier without adding too much to the distraction from the road ahead. Switches for other accessories should be grouped according to their function and clearly labelled; lighting switches should all be together, heated rear window and electric washer switches should be closely grouped.

There are many types of ready-made custom facias providing facilities for fitting additional instruments and accessory switches. The customiser may find that one of these will be a cheap convenient solution to the problem of making an attractive driving position. Switches and instruments can also be mounted on small panels fixed above or below the existing facia. Consoles of the type that fit on the centre of the facia or over the transmission tunnel are another good way of fixing a radio, tape player or accessory switches.

7:4 Anti-theft devices

The car is a vulnerable target for thieves and the added attraction of a tuned engine, radio, tape-player and other removable performance accessories increases the chance of a break-in.

The simplest electrical anti-theft device is the ignition immobiliser. This consists of a wire from the ignition coil contact to a switch well concealed under the facia. A short link wire earths the other terminal of the switch.

The effect of this device is to short circuit the contact-breaker points and render them ineffective. It is a measure that will defeat few determined and experienced thieves but it is a useful first line of defence against the joy-rider.

Better protection is afforded by fitting an alarm unit that operates the car's horn or, even better, an independent siren if the car is disturbed while parked.

A typical alarm unit consists of a vibration sensitive pendulum or spring that if set into motion operates a relay which switches on the horn. The relay is of the type that maintains current flow even if the stimulating current from the sensor contacts is discontinued. To prevent battery drain there is usually a bi-metal strip timing device in series with the relay which switches off the alarm after a pre-determined period. The alarm system is activated by a key switch located in such a way that the terminals are inaccessible from outside the locked vehicle.

Individual fitting instructions for alarm kits vary but the following rules should be observed to ensure efficient operation and the provision of maximum protection:

1 Tune the sensitivity of the sensor unit very carefully. Some are very temperamental and cannot distinguish between a cat landing on the bonnet, vibration from passing lorries and wind disturbance. To avoid causing a nuisance while tuning the unit connect a test lamp across the terminals in place of the horn.

2 Set the time control for at least 25 seconds but not more than 45 second of horn blast.

3 Site the lock switch in as well concealed a position as possible. A good place is inside the petrol filler flap provided that the tank is sealed with a separate filler cap.
4 Wire additional switch units on the door pillars and boot and bonnet surrounds. Ordinary courtesy lights switches could be used and they should be connected in parallel with the vibration sensing unit so that any or all of the thief's actions will set off the alarm.
5 On most cars the horn wires or terminals are readily accessible from outside the car. Ensure that the thief cannot disconnect the horn by either fitting an enclosure around the rear of the horn or wiring the alarm unit to a separate horn or siren placed in an inaccessible position.
6 Always remember to switch the device on when the car is left unattended.

7:5 Auxiliary lighting

Foglamps, spotlamps and reversing lamps are an essential addition to the rally car to increase safe road or special stage speeds and ease the task of driving in foul weather.

Foglamps:

A foglamps must be mounted low down on the front of the car so that the sharp cut-off at the top of the light beam is well below driver's eye level. To be fully effective a pair of foglamps should be fitted. A pair can be used with the headlights extinguished in fog and driving snow according to the law. But the lamps must be fitted within the areas defined by the law for headlamps (see legal requirements – **Chapter 8**).

A single foglamp cannot be used on its own and its performance will be limited by the light scattered from the dipped headlamps.

There are two good positions to mount foglamps, either on the bumper or front apron or on special brackets (obtainable from accessory shops) which are secured to the front grille. If possible avoid fitting lamps below the front bumper where they are likely to be damaged in parking or by flying stones.

Most lamps are supplied with a fitting kit containing the correct number of bullet connectors, a switch, and sufficient wire of the correct current carrying grade. If these components are not supplied ensure that the wire used for fitting is suitable for carrying a continuous current equal to the demand of both lamps. If they are 60 watt units the total power will be 120 watts and the current will be 10 amps.

Wiring should be taped neatly to the car's existing loom and fitted to switches and lamp with proper connectors. Apply additional protection to connectors behind the front grille by wrapping each one with plastics tape to prevent the ingress of dirt and moisture. The light switch should be firmly secured to the facia on a subsidiary switch panel or in a hole out into a free part of the instrument panel.

Connect the supply lead from the switch to the battery auxiliary fuse box terminal. But bear in mind that current load of up to 10 amps might overload this fuse and it may be necessary to fit a relief fuse box or a relay circuit. It is a wise precaution to connect a small warning lamp in parallel with the lead to the lamps, particularly if the lamps have protective covers fitted when not in use.

Spotlamps:

As a spotlamp is fitted to extend the range of the car's normal lighting it is realistic to fit only one of these units. A single spotlamp can be fitted anywhere at the front of the car provided it is not more than $3\frac{1}{2}$ ft from the ground and below 2 ft from the ground. The best position for the lamp is probably at the offside of the car with its beam directed towards the nearside to pick out the kerb. Wiring is the same as that for foglamps – long range driving lamps can be fitted in the same way. Special bars for lamp mounting are available for most cars. The mounting must be firm; a wavering beam is very distracting.

CHAPTER 8

Legal requirements and insurance

8:1 Introduction
8:2 The car's documents
8:3 Horns
8:4 Speedometer
8:5 Legal lighting requirements

8:6 Noise
8:7 Insurance grouping
8:8 Modifications that would invalidate an existing policy
8:9 Insurance for competition

8:1 Introduction

Cars figure large in the statute books and are covered in various sets of legislation. Chief among these are the Road Traffic Acts, the Road Vehicles (Registration and Lighting) Regulations, the Construction and Use Regulations and the Road Vehicles Lighting Regulations.

The Construction and Use Regulations are the ones that have the greatest relevance to the home car constructor. These useful but extremely complicated rules for vehicles need not affect the performance tuner concerned with a vehicle prepared for the road, provided that all the car's original equipment is in good order and working. Particular attention should be given to the brakes, steering and tyres.

Problems usually arise when the body of the car is modified. Sharp projections on the car may be considered to be illegal under the provision of the Construction and Use Regulations which demands that all parts and accessories on the vehicle must be in such a condition that there is no danger to any person in the vehicle or on the road. Sharp front or rear seams, lamp brackets and projecting exhaust pipes and sharp bonnet mascots are all frequently cited as transgressing the law. Road wheels must be covered by wings or mudguards.

It should also be borne in mind that a car in a dangerous and unroadworthy condition may be considered by an insurance company to have invalidated the insurance.

Specific regulations covering some of the points of law most home mechanics have to contend with are covered in the following sections.

8:2 The car's documents

The car's registration book (or log book) contains the following details: make, model, colour, chassis number, engine number and engine capacity. Much of this information could change during the process of tuning or renovating a car.

All changes should be detailed in writing to the licensing authority or the DoE's Driver and Vehicle Licensing Centre at Swansea (if the Registration Book has been surrendered and a computerised Registration Document issued). The notification of changes should be accompanied by the current Registration document.

Provided that the changes are of a fairly minor nature, such as raised capacity, changed engine number, complete respray, the change to the log book will be quickly implemented. Renewal of the chassis and a complete rebuild has usually meant re-registration. In a few instances, registration has been transferred with the substantial components of the original car used in the construction of a new vehicle, for example engine and transmission, although these exceptions have mainly been made with vintage cars.

In the case of a complete build it will be necessary for a car to be inspected by a DoE agent or examiner for compliance with the Construction and Use Regulations.

The computerisation of the records at the DoE's Swansea centre allows rapid inspection of registration particulars. It is therefore, easy for them to identify the use of engines or bodies from stolen cars or the falsification of chassis and engine number particulars to hide the use of stolen parts. Be particularly careful about the source of engines or car bodies bought from scrapyards. The majority of dealers are reputable, but it is an unfortunate fact that many stolen cars are processed through a few crooked scrapyard operators. Take care to keep a record of the source of all components used in modification and obtain receipts for all purchases.

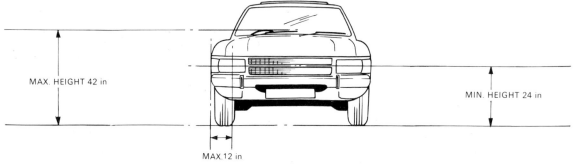

MAX. HEIGHT 42 in

MIN. HEIGHT 24 in

MAX. 12 in

FIG 8:1　Requirements for front lamps

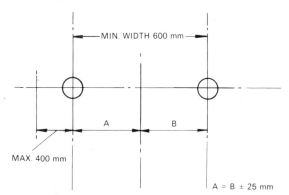

MIN. WIDTH 600 mm

A　　B

MAX. 400 mm

A = B ± 25 mm

FIG 8:2　Additional headlamp requirements

8:3 Horns

It is a legal requirement for a car to have an audible warning of its approach and location. The electric horn is the most practical way of fulfilling this requirement.

The horn's sound must be continuous, that is, the use of horns (particularly air horns) emitting two or more tones is restricted to the simultaneous emission of the two sounds. Horns that play tunes are forbidden.

The audible warning on a car must not sound like gongs, bells and sirens, or the two-tone horns used by police or emergency vehicles.

The use of horns is banned from 23.00 hours until 07.00 hours in a built-up area and when the car is stationary. This restriction does not apply when the horn is used as an anti-theft alarm.

8:4 Speedometer

The only instrument that the law requires on a car is a speedometer. It has to be accurate within 10 per cent of the car's velocity at speeds over 10 mph. Bear in mind that there may already be considerable inaccuracy in the speedometer, as manufacturers are known to fit optimistic speedometers that flatter the car's top speed and keep the driver within the law. Changes to the aspect ratio of the tyres or use of a different gearbox or differential may make considerable changes to the speedometer accuracy. Consult the instrument manufacturers for advice on

calibration if tests (see **Chapter 10**) show the accuracy to be outside the legal limits.

8:5 Legal lighting requirements

Lighting regulations are complex and subject to a great deal of change, particularly as and when European laws are adopted in Britain. In the guide which follows, the current position is given for vehicles used on or after January 1974. Vintage cars and other cars used before this date may be exempt from some of these provisions. Check with the legal advisers of a motoring organisation, the police or an advice bureau if you are contemplating making changes to a car's lights.

Front lamps: The law requires every vehicle to have two front lamps. They must be mounted at the same height and if over 7 watts in power the maximum height is 3 ft 6 inches. The centre of each light must not be more than 12 inches from the vehicle's widest point.

These provisions particularly relate to combined headlamp and sidelight units on modern cars.

Headlamps: For headlamps over 7 watts in power the above rule of 3 ft 6 inches maximum height applies with the addition that they must not have centres lower than 2 ft from the ground. The dipped beam must not dazzle at a height of 3 ft 6 inches and a distance of 25 ft. There is no legal obligation to have a left dipping beam, provided it does not infringe the anti-dazzle requirement.

As well as complying with the requirements for front lamps, headlamps must also be more than 600 mm apart and symmetrically placed about the car's centre line to within 25 mm. The outer edge of the headlamps must not be more than 400 mm in from the vehicle's widest point.

The headlamps must be a matched pair in shape, size and colour of light. Only white or yellow light is permitted. It is not permissible to have the light at each side under the control of separate switches. At least one bulb filament in each lamp must have a power over 30 watts. On a four-headlamp system, when on dipped beam, the outer pair must give the dipped beam and the inner lights extinguished.

Headlamps must be kept clean and when the vehicle is stationery they must be extinguished.

Obligatory rear lights: Two identical red lamps of at least 5 watts power must be mounted symmetrically about the car's centre line between a maximum height of 3 ft 6 inches and a minimum height of 15 inches. They

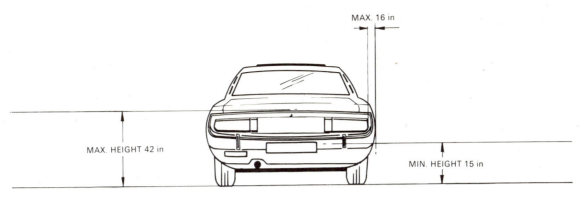

FIG 8:3   Requirements for rear lamps

must not be more than 16 inches from the vehicle's widest point. Two red reflectors must be fitted within the same height dimensions as the rear lamps.

Direction indicators: The only significant parts of the direction indicator legislation relevant to the car electrician are that the lights must be amber at the front and rear. Each lamp must have a power of between 15 and 36 watts, and the rate of flashing must be between 60 and 120 times a minute.

Direction indicators must at all times be maintained in a clean condition and in efficient working order.

The use of hazard warning flashers is permitted only when the vehicle is stationery as a result of a breakdown, accident or other emergency or when the vehicle is parked in a busy thoroughfare for the sole purpose of loading or unloading goods.

Stop lamps: Two red stop lamps of a power between 15 and 36 watts must be fitted a minimum of 600 mm apart, not lower than 400 mm from the ground and not higher than 1500 mm, at the rear of the vehicle.

Foglamps and spotlamps: Auxiliary lamps fitted below 2 ft from the ground can only be used in fog or falling snow. They can only be used in place of headlights if they comply with the headlight regulations. Maximum height above ground for auxiliary lights is 3 ft 6 inches. There is no necessity to fit auxiliary lamps in matched pairs. However, single lamps can only be used in conjunction with the car's headlights.

Reversing lights: No more than two reversing lights are allowed and the maximum power for each is 24 watts. The lights must be operated by a gearbox switch or panel switch, in which case a facia warning light must be used to warn when the reverse light is on. Lights must not dazzle an observer at 3 ft 6 inches height and 25 ft distance. The lights must only be used for reversing.

Registration plate light: A light capable of illuminating the rear number plate (maximum power 7 watts) must be fitted in such a way that no part of the lamp is visible from the rear – only the reflected light must be seen.

8:6   Noise

The greatest danger in modifying an exhaust system for performance requirements is that legal noise limits will be exceeded.

The law requires all vehicles to be fitted with a silencer or expansion chamber maintained in good working

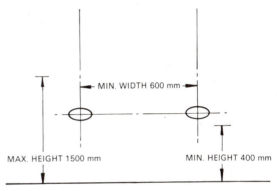

FIG 8:4   Stop lamp requirements

order. It must not be modified to increase exhaust noise.

Under the Construction and Use Regulations cars must not emit sound levels of over 88 dBA determined in a test procedure laid down by the British Standards Institution. It would be impractical to suggest home tuners had their cars tested to this procedure. The best advice is to demonstrate the car's acceleration from 30 to 60 mph to a panel of friends. Try and assess whether the sound produced is excessive compared to a new car performing the same test. New cars should emit a lower noise level than 88 dBA under EEC Directive (70/157/EEC).

8:7   Insurance grouping

In recent years the business of motor insurance has become considerably simplified, mainly because it is now a highly competitive activity. The small margins of profitable operation have shown up the weaknesses of deviating from the policies taken by large companies. Deviation has caused the failure of several insurers with revolutionary ideas on policy structure and price. Broadly, the choice is now between a policy with a member company of the Motor Conference or with a Lloyds syndicate.

Both groups of insurers base the premium charged for a particular model of car on a number of well defined factors. Principal among these are; the capacity and power of the engine; an engineer's assessment of the primary and secondary safety characteristics of the vehicle (handling, braking, roadholding, interior padding);

a judgment of the class of person to which the car appeals (an estimate of the risk inherent in the class of driver attracted to the car) ; the car's repairability in terms of the cost and availability of spare parts and the labour and time taken to repair the car; the insured's location.

The result is a group classification and an area rating. A Mini 850 is in Group 1, a Rolls-Royce in Group 7. High rates are charged for Inner London, and low for Inverness.

8:8 Modifications that would invalidate an existing policy

Modifying parts of the car that would alter its status under any one of these grouping considerations, without notifying the company concerned and accepting their reassessment, automatically invalidates the insurance policy. Examples of the types of modification that would invalidate the policy would be, for instance, the fitting a performance carburetter and using glass fibre body parts. Perhaps, rightly, the insurers would argue that the car's safety characteristics had been changed for the worse.

What kind of difference do tuning modifications make to the grouping ? The easiest example is to take a car that exists in various tuned and untuned forms, such as the Allegro. The standard 1100 Allegro is in Group 2 because of its low power and all round sensibility. It is a car that would appeal to the family motorist with responsibilities and therefore not a driver to take undue risks. Successively higher powered versions of the Allegro are in higher insurance groups. The 1750 Sport and HL are Group 5 – this makes a difference of about £90 in the basic premium. The reasons for the difference are plain. More power means a potential for more disastrous accidents and there is also the possibility that the man who outlays money on a small performance saloon instead of a larger family saloon is the kind of person who would run more risks in his driving.

How does this situation apply to the home tuner? In practice it means that whatever modification the tuner carries out to increase the power (and that could mean economy tuning measures that give a potential power increase) the insurer should be notified.

Almost certainly any increase in power above that permitted within a grouping category will be penalised by a rise in group rating. However, it is well worth telling the insurer about any other modifications made to the car's primary safety characteristics as well as those made to the power output. An insurer will appreciate that brake improvements and modifications to the handling and roadholding have been made to cope with the extra power. It may not stop the insurer raising the car one group rating but it might prevent him raising it two or three groups.

One of the queries uppermost in the insurer's mind will be why the tuning has been carried out. He will consider that to double the power of an 850 Mini is not an economy measure and that perhaps the owner is considering racing the car – perhaps on the streets. This immediately puts the owner into a high risk group as far as the insurer is concerned. If details of more modest power gains are being put to the insurer it is worth stressing the reason, particularly if they are genuinely aligned to economy motives.

The tuner's best course of action is to make a full explanation of the circumstances leading to the changes in the car's specification and the exact changes that have been made.

Should the rates quoted by the regular insurer appear to be punitive it is certainly worth consulting an insurance broker who will have a great deal of information available about insurers who are prepared to deal with the special risks presented by a tuned car. Do not be tempted by cut-price offers of insurance; make sure the insurer is a member of the British Insurance Association.

8:9 Insurance for competition

All insurance policies for normal road use exclude the entry of cars into any form of race, rally or speed trial. Special insurance cover has to be taken out for the period of the event. On large rallies and racing events the organisers will very often give the name of an insurer offering special rates for the particular event. For smaller events, brokers may be able to find an insurer willing to take on the risk or a special quotation can be obtained from the car's usual insurer. It is worth bearing in mind that especially high premiums are usually applied to tuned cars taken abroad.

CHAPTER 9

Tuning and new car warranties

9:1 New car warranties
9:2 Additional protection by legislation

9:3 Secondhand car warranties
9:4 Warranty schemes and their coverage

9:1 New car warranties

New cars are sold with a warranty or guarantee covering certain aspects of the vehicle's operation for a stated number of miles or for a specified time. The four top British manufacturers offer a warranty of 12,000 miles or one year, whichever comes the sooner. Foreign cars usually have a six month unlimited mileage or 6000 mile warranty period.

Items covered and those specifically excluded by the warranty vary from manufacturer to manufacturer, but generally replacement of faulty parts and no labour charge for the replacement is promised. It is usual to exclude tyres, batteries and often glass from the warranty as these may be covered by a component manufacturer's own guarantee. Some manufacturers, such as Renault and Volkswagen, include a sliding scale agreement in which a proportion of the costs of replacing parts, that can be proved to have failed due to faulty workmanship or materials, may be paid outside the warranty period. In Renault's case this agreement extends up to two years from new.

9:2 Additional protection by legislation

Additional protection to the new car buyer is provided by the Supply of Goods (Implied Terms) Act which came into force on May 18th, 1973. Although much vaunted as a piece of watertight legislation that fully restores the customers rights to redress for faulty goods under common law, despite any specific invalidations in a guarantee or warranty, it really does no more than offer the buyer a lever in the courts. It is certainly not the case that a buyer will be able to enforce any rights under the Act without entering into the full legal procedure of providing substantial proof of the defect's origin at the factory or in the salesroom. Theoretically the Act allows a car buyer to claim some recompense for early failure of parts up to three years from purchase.

The problem with both these methods of redress for faults on a car is that home servicing, tuning and the addition of non-standard parts to a car will certainly invalidate a warranty and will considerably weaken any case brought against a dealer or manufacturer.

A condition of the warranty given by most manufacturers is that the car must be serviced by their own accredited agents, that the car must not be modified in any way and that it must not be entered for any competitions. In practice this means that the home mechanic cannot lay a finger on the car until the warranty has run out, because if the garage can prove that any work has been done on the car at all – and there is very little that can be done on a car that an experienced mechanic can not spot – the warranty as a whole or part will be invalidated. That means no home tweaking of the car-buretter, ignition or valve gear.

If a buyer brought a case against a dealer under the Sale of Goods Act he would be bound to prove that the defect in question could not possibly have arisen as the result of any work that he had done on the car. In the case of engine components this would be extremely difficult to accomplish as even the most insignificant adjustment can have far reaching effects on the wear of stressed parts. Equally, it is almost impossible to establish whether a part became defective as a result of manufacturing weakness or as a result of less than fair wear and tear after the period of the warranty.

There are aspects of this situation, like the exclusion of the possibility of making tuning adjustments solely to restore the car to correct manufacturer's specification, that are extremely unfair to the new car owner. However, there is every justification for an exclusion on carrying out engine modifications and fitting tuning equipment. A mistake in doing this could cause considerable damage and the dealer could not be held responsible for the effects even if it could be proved that a manufacturing fault was a contributory cause of the disaster.

Manufacturers have become increasingly aware that more and more motorists want to make economy and performance modifications to their cars from new. And they are ready to admit that some tuning methods improve their cars. British Leyland were the first manufacturers to extend a full warranty when a dealer fits approved tuning equipment on new cars. Approved tuning equipment is that supplied by subsidiaries of the manufacturers, like Leyland ST and Dealer Team Vauxhall. The performance parts must be fitted by a franchised dealer for the warranty to remain valid.

9:3 Secondhand car warranties

Warranties and guarantees on secondhand cars are varied in their mileage, period and inclusions and exclusions. A car that is still under the new car warranty will be handed on with the remaining portion of the warranty still in force. This may be increased to the dealer's usual secondhand warranty distance or time if the remaining portion is shorter than this scheme. Dealers holding franchises for major manufacturers may opt to operate the secondhand guarantee scheme backed by that of the manufacturer, such as the BLMC Gauntlet scheme and Ford A1 guarantee. These generally offer a minimum of three months or 3000 miles, parts and labour guarantee and are only extended to cars under three-years old.

Individual dealer schemes may offer a period longer than this minimum and may cover parts only. The same exclusions on competition use and modifications as in a new car warranty usually apply. However, depending on the relationship that a customer has with his dealer and the trading conditions at the time of sale a dealer may respond favourably to a suggestion that a new part provided under a secondhand car warranty could be fitted by the owner. This arrangement could be advantageous to the owner in allowing other work to be carried out at the same time as fitting the part.

It is well worth asking a dealer if he will allow a particular modification to be carried out on a new or used car, but remember to obtain a written agreement that this will not infringe any aspect of the warranty.

9:4 Warranty schemes and their coverage

BLMC:

The warranty (Owner's Service Statement) lasts for 12,000 miles or 12 months and excludes tyres, batteries and the glass in lamps and windows.

Chrysler UK:

The Chrysler Protection Plan covers 12,000 miles or 12 months and additional ten-monthly diagnostic checks are given free after one year. Tyres and glass are excluded.

Datsun:

The scheme covers the car for 12,000 miles or 12 months, excluding tyres and glass and giving 6 months cover to batteries with partial compensation up to 12 months.

Fiat:

Unlimited mileage and 6 months cover offered by a warranty that includes the provision of two years anti-rust guarantee provided that the customer pays for two additional treatments at an authorised dealer.

Ford:

The Ford Customer Assurance warranty covers 12,000 miles or 12 months excluding certain body parts not supplied by Ford, and tyres. Batteries and radios are covered for 12 months irrespective of mileage and further proportional compensation will be given for batteries up to two years old. Free diagnostic checks are given at 15,000 and 27,000 miles.

Renault:

Six months unlimited mileage is covered by the Renault scheme which excludes windscreens and tyres but gives 12 month cover on the battery with a proportion of costs up to two years.

Vauxhall:

The 12,000 miles or 12 months warranty excludes tyres and batteries.

Volkswagen:

Warranty officially 6000 miles or 6 months but the company have announced a 'goodwill' policy of offering full or partial reimbursement for some failures outside the warranty period at the discretion of the company – glass is covered for 1 month or 600 miles.

CHAPTER 10

Testing a tuned car

10:1 Calibrating the car's instruments
10:2 Checking fuel consumption
10:3 Performance testing

10:4 Simple comparative test of power under load
10:5 Brake testing
10:6 Dynamometer and diagnostic testing

The effect of the many adjustments and tuning methods described in this book depend to some extent on experimentation whilst keeping a close check on the results of each step in the tuning process. The checks on performance and economy that can readily be made by the home tuner will not be quite as detailed as those published by Motor, Autocar, the Automobile Association or the performance car publications. Nevertheless, using the methods detailed below it will be possible to obtain reasonably accurate and reproducable results that can be compared with more authoritative figures to ascertain tuning progress. Keen performance tuners may wish to take performance testing a stage further by using a dynamometer installation. The sort of facilities offered at a typical electronic diagnosis and dynamometer bay are described at the end of this section.

10:1 Calibrating the car's instruments

To obtain accurate figures that are comparable with published testing results it is necessary to calibrate the cars instruments accurately. Professional road testers use a fifth wheel attached to the rear of the car. The unit, resembling a bicycle wheel, is geared to drive a very accurate alternating current generating head. The current produced is fed to an accurately calibrated AC voltmeter instrument designed to show road speed. The reason for such care is that the speedometer fitted to most production cars is usually extremely inaccurate – the law permits a ten per cent inaccuracy in the instrument and there are plenty of examples of readings that are worse than this.

Car speedometers are magnetic devices driven by cable from the transmission. Changes in wheel and tyre size and the gear ratio of the final drive (top gear and differential) will mean speedometer inaccuracy. The best way to overcome this is to consult the instrument manufacturers who will almost certainly be able to supply a new speedometer or a new drive gear to restore the calibration to within the usual limits of accuracy.

Check the speedometer by driving at a steady indicated speed of, say, 30 mile/hr over a measured distance, for example, between the mile posts on a motorway, and use a stopwatch to take the time. Use the formula below to calculate the actual speed of the car:

$$\text{Speed (mile/hr)} = \frac{\text{Distance}}{\text{Time}} =$$

$$\frac{\text{Distance between marks (miles)} \times 3600}{\text{Stopwatch time (seconds)}}$$

To eliminate error as far as possible repeat the distance and time measurements at 30 mile/hr (indicated) at least four times and use the average time for the calculation. The result may turn out to be in the region of 30 mile/hr, say, 31 mile/hr. At this low speed you can assume the car is therefore travelling a true 30 mile/hr when the dial registers 29 mile/hr. However the timing must be repeated at 40, 50, 60 and 70 mile/hr. At 70 mile/hr (indicated) it is quite possible that the true speed may be 75 mile/hr and it will no longer be possible to assume that at an indicated 65 mile/hr the car will be travelling a true 70 mile/hr. The test must be repeated at the 65, 66 and 67 mile/hr indicated speeds to find which is the accurate mark for a true 70 mile/hr. Mark the dial of the speedometer (or use an overlay on the glass) with the true speed marks you have found.

The same factors which affect the speedometer reading also cause mileometer (odometer) inaccuracy. An

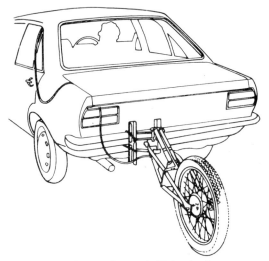

FIG 10:1 Professional tester's fifth wheel

additional problem is the state of wear and inflation of the tyres which can have a considerable cumulative affect over long distances.

The mileometer accuracy (often as much as 5 per cent adrift) can be checked on a journey over a known distance of at least 10 miles. A local authority surveyors department may be able to provide you with an accurate measure of the distance between two readily identifiable road marks, for instance between two bridges on a motorway. Accurate mileometer calibration is essential for comparing fuel consumption figures with those published in the motoring press.

A car's tachometer can only be calibrated by comparison with a known accurate instrument – ask a garage to check it out on the electronic diagnostic equipment.

10:2 Checking fuel consumption

The method of driving a car has by far the greatest effect on the car's fuel consumption therefore the technique of fuel consumption measurement (described later) can only be used with consistent accuracy if a repeatable cycle of speeds, accelerations, gear changes, braking and stoppages can be established over a reasonable distance. It follows, therefore, that the journey used for testing purposes should be over the same route for the same duration of time and in the same traffic conditions each time a test is made.

This is not impossible to achieve. A regular commuter run made at the same time every day over the same distance may be quite suitable. Traffic conditions will not vary significantly from day to day at the same hour. Weather conditions will, however, and it is best to restrict testing to calm days. It is said that the amount of water vapour in the air will affect figures and certainly the air temperature will, so the days chosen should be those with clear visibility and of similar temperature.

An alternative to the use of a fixed route is to take the overall fuel consumption over a week of motoring regardless of mileage but provided that week to week motoring conditions are similar. The same balance must be observed between town, open road and motorway.

Accurate measurement of the petrol consumed is also an essential part of mile/gall measurement. Find a petrol station that has pumps which display the quantity delivered to at least one decimal part of a gallon. Some modern pumps now have digital dials showing the amount delivered to two decimal places and these are the best type for accurate fuel measurement.

Always fill the car up at the same pump with the car facing in the same direction and with the same load in the car. This ensures that the level of the car remains the same. Fill the tank right to the brim or at least to a visible point in the filler neck. Rock the car during filling of the last gallon or so to ensure that air bubbles and traps in the tank are released – it is surprising how much extra the tank will take on some cars when this rule is observed.

The formula to calculate mile/gall takes into account the correction factor determined in checking the accuracy of the mileometer.

$$\text{Correction factor} = \frac{\text{Actual miles travelled}}{\text{Mileometer miles indicated}}$$

Therefore:

Fuel consumption (mile/gall)

$$= \frac{\text{Indicated mileometer distance} \times \text{Correction factor}}{\text{Number of gallons used}}$$

In comparing the figures obtained by the method above, some subjective judgment of the overall conditions encountered during the test run must be made. For instance it is usual for a published road test to quote constant speed fuel consumption figures, overall consumption and touring consumption. Ignore the constant speed fuel consumption figures; these are misleading and although they are often used by manufacturers in advertising they are meaningless in day to day motoring. The overall fuel consumption given in road tests corresponds to a mixture of usage which includes motorway, town and country touring use. This may correspond very well to the type of driving used in a regular commuter run. However a test taken on country and A-roads at moderate speeds would correspond more to the touring consumption quoted in road tests.

10:3 Performance testing

Two performance figures widely quoted in road tests are useful measures of the car's capabilities – the 0-50 mile/hr time (through the gears) and the 30-50 mile/hr time (in top gear only).

Ideally, testing of this kind should be carried out on a private track. The nearest to this that the home tuner may be able to find is a disused airfield. However a straight, more or less flat, stretch of road about $\frac{3}{4}$ mile long will do provided that it is free of traffic at the time of testing. This may necessitate an early morning start. A colleague will be required to operate the stopwatch.

Before performance testing run the car for several miles to warm the engine up to normal operating temperature. Decide the maximum engine rev/min to be attained in each gear and if a comparison between test and published figures is sought check the published maximum rev/min used for the specific test. Bring the car to rest and accelerate up to about 3000 rev/min and activate the stopwatch as soon as the clutch is engaged. Allow the engine to accelerate to the set maximum revolutions before each gear change and give the timekeeper a clear signal when the accurate 50 mile/hr is reached. Repeat the test several times in each direction to take into account differences in wind direction, traction at the start point, road levels etc. Average the results obtained to give the test figure.

It will be appreciated that this procedure will be heavy on the car's tyres (wheel slip will almost certainly be induced at take-off), transmission and engine. To avoid some of the strain it may be more suitable to take the time between 30-50 mile/hr in the gears in which case the test would consist of accelerating hard from say 10 mile/hr in first gear or second (depending on the maximum speed in each gear) and proceeding as above until 50 mile/hr is reached.

The 30-50 mile/hr time in top gear is taken by engaging top gear at below 30 mile/hr, say 27 mile/hr, flooring the accelerater and timing the interval between attaining the true 30 mile/hr and true 50 mile/hr. The measurement obtained is related to the power under load. However for some other purposes the time is not required – indeed an even heavier load can be advantageous to show up minute performance differences.

10:4 Simple comparative test of power under load

Carburetter, ignition and plug gap setting can be checked out very simply by acceleration up hill in top gear. A hill should be chosen of moderate gradient with a straight approach sufficient to allow safe acceleration to 30 mile/hr. Set out two marks on the hill about 200-300 yards apart. Accelerate to pass the bottom mark at 30 mile/hr in top gear with the accelerater floored. Proceed up the hill at full throttle in top gear and note the speed registered when the top mark is reached. Make whatever minor adjustments are being tested – remember that only one type of adjustment should be made at a time. Repeat the test aiming for the fastest speed at the top mark.

It is worth repeating here the dangers of over-advancing the spark and over-weakening or enriching the mixture. Dire engine troubles can result.

The simple test described above is the best way of carrying out ignition retardation to tune an engine to accept a slightly lower grade of fuel. The heavy load placed upon the engine will cause intensive pinking if the ignition is too advanced. The test should be repeated with the lower grade of fuel until the pinking can no longer be heard.

10:5 Brake testing

It will only be necessary to test brakes if changes have been made to the hydraulic system (a servo has been fitted, for example), to the pad or lining material, to the drums or to the front/rear balance. Again this is a comparative test to be made before and after any changes are made to the braking system.

On a straight secluded stretch of road find or make a distinctive braking point on the road side. Approach the point at 30 mile/hr and as close as possible to the marker make a full emergency stop. Measure the braking distance with a surveyor's tape measure.

The procedure above is a very rough guide to braking performance. Normally tests are carried out with some more complex equipment. The Tapley meter is the widest used brake testing instrument. One pattern can be obtained quite cheaply from the manufacturers (see FIG 10:2). Another type consists of a manometer filled with coloured liquid. A scale is set against the manometer, calibrated from 0 (the liquid level at rest on a flat road) to 100 and the figures refer to percentage efficiency; 100 per cent represents a deceleration of 1g.

The meter is stuck to the facia or windscreen of the car and levelled at rest. The brake test, full stop from 30 mile/hr, is carried out and the lowest point to which the liquid level dips is noted. Most modern cars achieve efficiency readings in excess of 95 per cent, that is, a stop of 0.95g deceleration.

Brake fade is an important factor to be considered in stopping performance. In the simple test a measure of brake fade would be provided by the distance taken to stop after about 15 full stop applications of the brakes from 30 mile/hr, using approximately the same pressure for each application. More definitive is the readout of the pedal pressure required to produce a stop of 0.5g from 30 mile/hr at first application and at the 15th application (see **Chapter 6** for measures to combat fade).

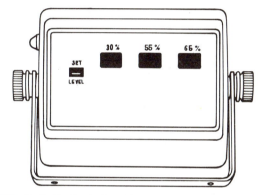

FIG 10:2 Tapley brake efficiency indicator

10:6 Dynamometer and diagnostic testing

An easier and more precise, but unfortunately more expensive means of obtaining a great deal more information about the car's performance is to use an instrumented test bay and dynamometer (or rolling road).

There are testing stations with Sun, Crypton and other types of equipment in all major towns. The cost of using the facilities varies with the range of tests the stations can offer but a rough guide is between £9 and £15 an hour. This is steep but a great deal of testing can be done in a fraction of this time.

Dynamometer testing:

A dynamometer, often called a rolling road, is an instrument that simulates road, inertia and wind resistance loads on a car while it stands still. The drive wheels of the car to be tested are located on rollers which are electromagnetically or hydraulically braked to produce an exact representation of the forces the driving wheels would have to overcome on the road. Variable loading on the rollers is used to preset the dynamometer for the car's weight and air drag.

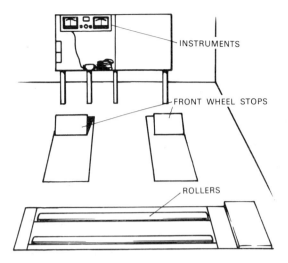

FIG 10:3 Rolling road dynamometer

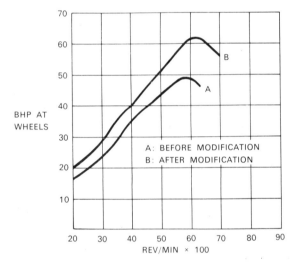

FIG 10:4 Typical 'before and after' power curves

The dynamometer is used to measure the power available at the wheels in brake horsepower (bhp). This is measured by revving the engine steadily at a fixed rev/min in top gear and balancing the electromagnetic braking effect applied to the rollers until the car can make no headway, that is cannot accelerate beyond that engine speed. The direct readout in bhp plotted against rev/min in graph form is the engine/transmission power curve.

The beauty of the dynamometer is that the effect of various adjustments can be seen immediately as the car can be tuned – while the rolling road is running if necessary. The same machine can be used to take acceleration times.

Some dynamometers have a facility for testing brakes. In this case the rollers are driven to rotate the car wheels and then the car's brakes are applied. The dynamometer measures the force that the brakes are capable of applying to arrest motion. Brake testing is usually carried out on a separate dynamometer however, and such an instrument is capable of reading out the different braking effort applied at each wheel.

Dynamometers are perhaps more difficult to find than the normal range of diagnostic equipment. More typical equipment will provide the following information about the engine's performance:

1 dwell angle of the contact breakers and distributor wear
2 plug firing voltages
3 plug condition
4 coil condition
5 high tension lead insulation condition
6 vacuum and centrifugal advance and stroboscopic timing
7 dynamo/alternator output voltage regulator and cut-out performance
8 starter motor cranking voltage
9 mixture setting by measurement of carbon monoxide in exhaust
10 valve timing.

Skilled interpretation of the instruments in the diagnostic set by an experienced operator can show up the slightest sign of poor performance or unsuspected wear. Costs for using equipment like this vary from £7.50 per session to £12 per hour. There are now mobile diagnostic tuners who will carry out this service at a motorist's own home or in the office car park. These operators advertise in the local newspapers for the area in which they hold a franchise.

Index

Figures in bold type refer to illustrations

A

Advance mechanisms 50
Aerofan **45**
Air cleaners 37
Air dams 99
Air filter 37
Air filter, home-made **37**
Allen keys 14
Amal carburetters, see carburetter
Anti-roll bar 87
Anti-theft devices 101
Anti-tramp bar **88**
Arch, wheel 100
Autolite carburetter, see carburetter
Axle stands **13**
Axle tramp 79

B

Balancing 44
Ball joint removers **18**, 19
Bodywork, modification 22
Bottle jack **13**
Brakes 89, 94
Brake booster 98
Brake disc 95
Brake disc assembly **95**
Brake disc assembly, sliding yoke .. **96**
Brake disc conversions 97
Brake drum 94
Brake drum assembly **95**
Brake drums, light alloy **97**
Brake efficiency indicator 111
Brake fade 94
Brake grabbing 94
Brake, handbrake 96
Brake hydraulic system modification .. 98
Brake lining **96**
Brake maintenance 94
Brake modification 96, 97
Brake performance 94
Brake, safety 98
Brake servo unit 98

Brake testing 111
Brake, uprated drums 97
Bump stop **88**
Burette 19

C

Calipers, vernier **16**
Cam, harmonic profile **43**
Cam, polydyne profile **43**
Camber, negative 92
Camber, positive 92
Camshaft 43
Capacitor 64
Carburation 23
Carburetter, Amal 33, **36**
Carburetter, Autolite, see FoMoCo
Carburetter, Dellorto **23**
Carburetter, FoMoCo 30
Carburetter, FoMoCo, tuning .. 30, 31
Carburetter, Fish, operation **36**
Carburetter, Fish, economy tuning .. 36
Carburetter, Fish, performance tuning .. 36
Carburetter, Fish (twin) **35**
Carburetter, Motorcraft, GPD, see FoMoCo
Carburetter, Minnow Fish, see Fish
Carburetter, Nikki 33, 33
Carburetter, Reece Fish, see Fish
Carburetter, Stromberg CD **27**, 28
Carburetter, Stromberg, CDSE air bleed screw **29**
Carburetter, Stromberg, checking float height **30**
Carburetter, Stromberg, cross section .. **29**
Carburetter, Stromberg, exploded diagram .. **28**
Carburetter, Stromberg, orifice disc starting
   device **29**
Carburetter, Stromberg, tuning .. 29, 30
Carburetter, SU 23, **24**
Carburetter, SU, cross section .. **24**
Carburetter, SU, exploded diagram .. **25**
Carburetter, SU, float level, checking .. **26**
Carburetter, SU, jet centring .. **26**
Carburetter, SU, needle fitting .. **24**
Carburetter, SU, piston **26**

Carburetter, SU, twin	**27**
Carburetter, SU, economy tuning	26
Carburetter, SU, performance tuning	26
Carburetter, Weber DCD, DFE, DFM	31
Carburetter, Weber DCD, float level	**33**
Carburetter, Weber DCD, tuning	33
Carburetter, Weber DCOE, exploded diagram	**34**
Carburetter, Weber DCOE, tuning	35
Carburetter, Weber DFE/DFM, exploded diagram	**32**
Centrifugal timing control	**52**
Circlip pliers	**15**
Circuit tester, home-made	**16**
Clutch, cable operation	71
Clutch, diaphragm spring	**72**
Clutch, maintenance and overhaul	71
Clutch, modifying	72
Clutch, operation, hydraulic	72
Clutch plate	72
Clutch, refitting	72
Clutch unit, uprated	**73**
Colortune kit	36, **37**
Combustion chamber, determining volume	38
Combustion chamber, grinding	**40**
Combustion chamber, measuring volume	**38**
Combustion chamber, modifying shape	**40**
Compression tester	**17**
Compression ratio	39
Compression ratios on flat-headed engines	43
Compressor, valve spring	**19**
Contact breaker	50
Contact breaker dwell angle	50, **52**
Contact set, Lucas one-piece	**57**
Contact set, Lucas two-piece	**57**
Cooling fan, electric	**44**
Cooling system	22, 44
Cooling system, economy tuning	44
Cooling system, performance tuning	44
Cylinder head gasket, determining volume	39
Cylinder head holding bolt, improvised	**40**
Cylinder head machining	42
Cylinder head measuring	38
Cylinder head skimming	42
Cylinder head modification	37, 38
Cylinder head preparing	38

D

Damper, lever arm cutaway view	**81**
Dampers	80, 81, **82**
Dampers, adjustable	**81**
Diagnostic testing	111
Dial gauge	**19**
Differential	73
Differential, limited slip	75
Differential swops	75
Distributor	49
Distributor, Bosch	59, **60**
Distributor cap defects	**57**
Distributor, capacitor	53, 57
Distributor, cutaway view	**51**
Distributor design variations	59
Distributor, Ducellier	59, **60**
Distributor, Hitachi	59, **61**
Distributor, lubricating	58
Distributor, Lucas Opus	**65**

Distributor, Marelli	59, **60**
Distributor, Motorcraft/Autolite (Ford)	**59**
Drum, uprated	97
Dynamometer, rolling road	111, **112**

E

Engines, modified	66
Engine swops, BLMC	22
Engine swops, Ford	22
Engine swops, Vauxhall	22
Exhaust systems	22
Exhaust system, tuned	**45**
Exhaust system tuning	45

F

Facias, customising	101
Feeler gauges	**15**
Final drive ratios	73
Fish carburetter, see carburetter	
Foglamps	102
FoMoCo carburetter, see carburetter	
Front lamps	104
Front suspension, see suspension	
Fuel consumption, checking	110

G

Gauges, feeler	**15**
Gauge, tread depth	**17**
Gauge, tyre pressure	**17**
Gearbox, BLMC	**68**, 71
Gearbox, Chrysler	71
Gearbox, Ford	71
Gearbox maintenance	67
Gearbox modification	70
Gearbox overhaul	67
Gearbox remote control linkages	71
Gearbox selector mechanisms	71
Gearbox, typical	**68**
Gearbox, Vauxhall	71
Gears, close ratio	70
Glass fibre parts, fitting	99
Grips	15

H

Half shafts	75
Handbrake	96
Handling characteristics	78
Head, see cylinder head	
Headlamps	**104**, 104
Horns	104
Hydraulic damper	80

I

Ignition coil	53, 56
Ignition coil, cross section	**53**
Ignition system circuit	**50**
Ignition systems	49, 62
Ignition system, ballast resistor	54
Ignition system, BLMC all-electronic	65
Ignition system, capacitor discharge	64
Ignition system, contactless	64
Ignition system, Lucas Opus	65
Ignition system, Lumenition	**64**
Ignition system, transistor assisted	63
Ignition timing, static	59

Ignition timing, stroboscopic 60
Indicator, direction 105
Instruments 14
Instruments, calibrating 109
Insurance 103, 105, 106
Interior, customising 100

J
Jacks, bottle **13**
Jacks, trolley **13**
Jet setter device **27**

L
Leads 56
Legal protection, new cars 107
Legal requirements, general 103
Lightening 44
Lighting, auxiliary 102
Lighting, legal requirements 104
Lights, front lamps 104
Lights, headlamps 104
Lights, lamps 102, 105
Lights, rear lights 104, **105**
Lights, registration plate light 105
Lights, reversing light 105
Lights, spotlamps 102, 105
Lights, stoplamp **105**, 105
Lining, brakes **96**
Live axle suspension **79**
Lowering kit, rear axle **83**
Lubrication system 45

M
MacPherson strut suspension .. 82, **84**, **85**
Manifolds, matching to head 41
Micrometer 15, **16**
Motorcraft carburetter, see carburetter

N
Negative earth TAC circuit **63**
Nikki carburetter, see carburetter
Noise 105

O
Oil cooler **46**
Oversteer **78**

P
Panhard rod **88**
Performance testing 110
Piston crown volume measuring 39
Pistons, Ford **43**
Pistons, high performance 43
Pliers 15
Plug, see spark plug
Polarity, warning 56
Port face, marking gasket size **42**
Port, polishing tool **42**
Ports 41
Ports, matching 42
Positive earth TAC circuit 62
Power curves **112**
Propeller shaft **74**, 75
Puller, flywheel 18
Puller, hub and flywheel, BLMC **18**
Puller, universal 18

R
Race cars 11
Racing saloon 12
Rally cars 11
Rear axle **74**
Rear axle, lowering kit **83**
Rear suspension, see suspension
Registration 103
Reinforcement, metal 100
Release bearings 72
Rev counter 17
Road cars 11
Roll centre 79
Roll over bars 100

S
Saloon racing 12
Screwdrivers 15
Seating 100
Shock absorbers 80
Slip angle 78
Sockets 14
Spanners 14
Spanner, brake adjusting **14**
Spanner, plug **14**
Spark plug 54, **55**, 56
Spark plug, gap tool 15, **16**
Speedometer 22, 104
Spotlamps 102
Stands, axle **13**
Steering 79
Steering wheels 100
Stroboscopic timing lights **16**
Stromberg carburetter, see carburetter
Stud holes, wheel **90**
SU carburetter, see carburetter
Suspension 22
Suspension, BLMC hydrolastic 87
Suspension, coil spring, independent .. **84**
Suspension, front 84, 86
Suspension, front, torsion bar **86**
Suspension, front, with anti-roll bar .. **87**
Suspension, function of 77
Suspension, Hydragas 87
Suspension, Hydropneumatique 87
Suspension, live axle **79**
Suspension, lowering 83
Suspension, MacPherson strut **84**
Suspension modifications 77, 80
Suspension, rear, modifications 83
Suspension, rubber spring 87
Suspension stiffening 83
Suspension, trailing arm **82**
Suspension, wishbone 79, **82**
Suspension, wishbone independent .. **85**

T
TAC circuits 62, 63
Tachometer 17
Tappet 15
Tappet, SPQR adjuster **15**
Thermostat **44**
Throttle, steering 79
Thrust bearings 72

Timing card **60**
Timing, centrifugal **52**
Timing, static 59
Timing, stroboscopic 61
Timing, vacuum **52**
Tool, spark plug gap 15
Tools, care 14, 19
Tools, special 17
Tools, tappet adjusting 15
Tools, valve adjusting 15
Torque wrench **14**, 14
Torsion bar suspension **86**
Toughening 44
Transmission, components 75
Transmission, tuning 67
Tread, tyres 93
Tuning, capital costs 12
Tuning, definition 10
Tuning, direct costs 12
Tuning, indirect costs 12
Tuning, operations involved 10
Tuning, savings 12
Tuning, scope for 9
Turbocharger operation, principle of .. **47**
Tyre, pressure gauges **17**
Tyre, tread depth **17**
Tyres 89, 92
Tyres, care 93
Tyres, cross-ply **93**
Tyres, matching to wheels 93
Tyres, radial **93**
Tyres, rubber mix 93
Tyres, tuning for economy 93
Tyres, tread pattern 93
Tyres, wide 92

U
Understeer **78**
Universal joints 75
Universal puller 18

V
Vacuum timing control **52**
Valves 41
Valve, adjusting tools 15
Valve spring compressor **19**
Valve suction grinder **19**
Valve, modified 41
Valve, standard **41**
Valve seats 41, 41
Variable pitch fan **45**

W
Warranties, new car 107
Warranty schemes and coverage 108
Weber carburetter, see carburetter
Wedge ball joint remover **18**
Wheels 89
Wheels, alloy **91**
Wheels, arch flare 100
Wheels, attitude **91**
Wheels, construction 90
Wheels, flares 99
Wheels, lock up 94
Wheels, spacers **91**
Wheels, steel, cross section **90**
Wheels, steering 100
Wheels, stud holes, defining position .. **90**
Wheels, wide 91, 92
Wheels, wide rim steel **91**
Wishbone suspension **79**
Workshop practice 13

KNOW MORE ABOUT YOUR CAR

with an Autobook Workshop Manual

Includes detailed information about:

The Engine
The Fuel System
The Ignition System
The Cooling and Heating System

The Clutch
The Transmission
The Rear and Front Suspension
The Steering System
The Electrical System
The Bodywork

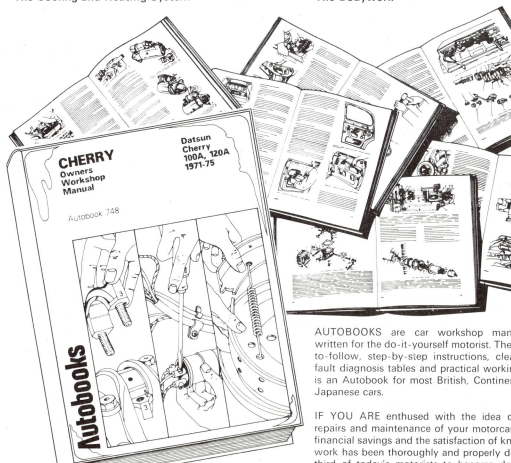

CHERRY
Owners
Workshop
Manual

Datsun
Cherry
100A, 120A
1971-75

Autobook 748

Autobooks

AUTOBOOKS are car workshop manuals specially written for the do-it-yourself motorist. They feature easy-to-follow, step-by-step instructions, clear illustrations, fault diagnosis tables and practical working hints. There is an Autobook for most British, Continental or popular Japanese cars.

IF YOU ARE enthused with the idea of tackling the repairs and maintenance of your motorcar, you will find financial savings and the satisfaction of knowing that the work has been thoroughly and properly done have led a third of today's motorists to become do-it-yourselfers. Many rely on an Autobook for detailed, accurate instructions.

AUTOBOOKS may be obtained from all good motor accessory shops or bookshops, or from Autobooks Ltd., Golden Lane, Brighton BN1 2QJ, telephone Brighton (0273) 721721.

Other titles in the Autocare series:

Electrical Systems – including Tapes and Radios

Bodywork Maintenance and Repairs – including interiors